KW-240-527

# Contents

7002111860

WITHDRAWN
FROM
UNIVERSITY OF PLYMOUTH
LIBRARY SERVICES

# SEVEN DAY LOAN

### This book is to be returned on
### or before the date stamped below

20. MAY 1998
CANCELLED

24 CANCELLED

CANCELLED
2 2 APR 1999

2 8 FEB 2000

8 MAR 2001

1 2 MAY 2003

2 7 OCT 2003

## UNIVERSITY OF PLYMOUTH

# PLYMOUTH LIBRARY

### Tel: (01752) 232323
This book is subject to recall if required by another reader
Books may be renewed by phone
CHARGES WILL BE MADE FOR OVERDUE BOOKS

70 0211186 0

TELEPEN

European Community
Environmental Policy
in Practice

# Volume 1

# Comparative Report:
# Water and Waste in Four Countries

**European Community
Environmental Policy
in Practice**

## Volume 1

# COMPARATIVE REPORT:
# WATER AND WASTE IN
# FOUR COUNTRIES

A Study of the Implementation
of the EEC Directives in France,
Germany, Netherlands and
United Kingdom

by

*Nigel Haigh*

with

*Graham Bennett
Pascale Kromarek
Thierry Lavoux*

Graham & Trotman

First published in 1986 by
Graham & Trotman Limited
Sterling House
66 Wilton Road
London SW1V 1DE

Graham & Trotman Inc.
13 Park Avenue
Gaithersburg
MD 20877
USA

© Institute for European
Environmental Policy, 1986

ISBN 0 86010 549 0

PLYMOUTH POLYTECHNIC
LIBRARY

Accn
No          211186 - 0

Class
No        344.046343HAI

Contl      0860105490

This publication is protected by international copyright laws. All rights reserved.
No part of this publication may be reproduced, stored in a retrieval system, or
transmitted in any form or by any means, electronic, mechanical, photocopying or
otherwise, without the prior permission of the publishers.

Typeset in Great Britain by the Castlefield Press, Wellingborough
Printed and bound in Great Britain

# PREFATORY NOTE

This report is one of a series of studies by the Institute for European Environmental Policy on the implementation of EEC Directives relating to water quality and waste disposal by the Member States of the European Community. National reports covering the Netherlands, France and the Federal Republic of Germany are being published in this series together with this comparative study. A similar report for Britain, but covering the implementation of all environmental Directives, has already been published ( Nigel Haigh, *EEC Environmental Policy and Britain: An Essay and a Handbook*, ENDS Ltd, 40 Bowling Green Lane, London ECIR ONE, 1984). Studies in the remaining Member States are in preparation.

This study was completed in March 1985 but was revised to include some developments up to the end of 1985.

# ACKNOWLEDGEMENT

This comparative report was financed by the UK Department of the Environment and is published with its agreement. The national reports on which it is based were financed from other sources. Its publication has been made possible by a grant from the Commission of the European Communities. The views expressed in the report are those of the authors: in no circumstances should they be taken as an authoritative statement of the views on the Commission.

# 1

# Introduction

## ASSESSING EC ENVIRONMENTAL POLICY

It goes without saying that the purpose of the environmental policy of the European Community (EC) is to improve the environment in Europe. In a national context it is well known that the connection between an item of environmental legislation and a measurable improvement is not always easy to demonstrate and with EC legislation the problem is that much more difficult. Not only is the chain between EC legislation and an actual improvement at least one link longer than in a national context — since EC Directives have to have been translated into national legislation or administrative measures before they can take effect — but a number of different Member States are involved each of which may be affected in different ways.

The analogy of a chain linking legislation to an environmental improvement is in fact not very appropriate in an EC context. A better analogy would be a spider's web with the EC institutions at the centre connected to the Member States at the circumference, and with the Member States connected to one another by strands, some of which pass through the EC institutions but some of which do not. Influences can flow along all these connecting strands, but at this point the analogy breaks down. This is because any movement at one point on the web is not necessarily felt immediately at others along the laterals in the same way as it is along the radials. There is no doubt that the influence of the EC institutions, at the centre, on the Member States, at the circumference, is very much stronger than the bilateral influences between the Member States. The reason for this is that the influence of the EC institutions stems from their ability to adopt legislation binding upon the Member States, whereas the influences between Member States are usually only the influences that come from example.

The environmental policy of the EC can in large measure be regarded as the sum of the items of this EC environmental legislation together with what is then done to implement that legislation. Any one item may have been inspired by an existing item of national legislation so the EC legislation may result in very little change in that country. The same item of EC legislation may nevertheless have a significant effect in other countries. Any attempt to assess the achievements of EC environmental policy must therefore involve three steps: first an examination of the legislative texts themselves to determine their intentions; second, a consideration of how each item of EC legislation has been translated into national legislative and administrative measures in all the countries; and third, an assessment of what has happened in practice. This must be done for each item of EC legislation in each country: there is no short cut.

# THE NATIONAL STUDIES

This comparative report on the impact in four countries of the EC Directives on water and waste is based on four self-contained studies carried out by four different people. It was no surprise to find that those who were interviewed in the course of the study in each of the four countries wanted to know how the same Directives were being implemented in the other countries. This comparative report is intended to fill that want.

The study of the United Kingdom, which was published in book form* in February 1984, was carried out by Nigel Haigh and was financed by the Nuffield Foundation and by the European Cultural Foundation. The UK study covered other fields as well as water and waste (air, chemicals, wildlife, noise) and took over three years. Some updating has therefore been necessary for the purposes of this comparative report. The methodology adopted for that study was described in the book and involved the distinction between *formal* compliance or implementation (the legal and administrative steps taken in response to a Directive) and *practical* compliance (what is done in practice). The book showed that the impact of EC policies can involve changes in ways of thinking about issues, and changes in the relations between central and local government, in addition to changes in legislation and practice.

The studies of Netherlands, France and Germany were carried out respectively by Graham Bennett, Thierry Lavoux and Pascale Kromarek

---

* *EEC Environmental Policy and Britain: An Essay and a Handbook*, published 1984 by Environmental Data Services Ltd (ENDS)

under a contract from the Commission of the European Communities begun in 1983 and finished in 1984. Financial contributions to these studies were made by the French Ministry of the Environment, the German Umweltbundesampt and the Netherlands Ministry of the Environment. Each of these three authors followed the method set out in the study of the United Kingdom although each worked largely independently of the others.

This comparative report was carried out under a contract from the British Department of the Environment. As well as being based on four self-contained studies, it is also the work of the four authors. Although the drafting of the section on each Directive to be found in Chapters 3 and 5 was in the first place the work of the principal author, each draft has been checked and added to by the other three authors. Each section provides a summary of the material to be found in the four national reports but is presented so that the most interesting differences can be compared. Anyone wanting further information about a particular country should therefore refer to the national reports. As often happens when comparisons are made between countries, what is a significant point in one country may not be significant in another. The national reports have therefore sometimes had to be supplemented, and some material appears in this comparative report that does not appear in the national reports. Material in the national reports — often collected in mid-1984 — has also sometimes had to be updated for this comparative report that was completed in March 1985.

# THE INSTITUTIONAL ARRANGEMENTS FOR ENVIRONMENTAL PROTECTION

One of the greatest difficulties with any report that compares a particular subject in different countries lies in relating that subject to the institutional arrangements for handling it. In a national context these institutional arrangements are nearly always taken for granted since the reader can be expected to be familiar with them. This familiarity cannot be assumed with a comparative report. For this reason, the following chapters which deal with the individual water and waste Directives are preceeded by chapters which briefly describe the legislation and responsibilities for water and waste in the four countries. Even those chapters themselves require to be set in the broader context of the constitutional arrangements for the machinery of government administration in the countries concerned.

All four countries discussed here are parliamentary democracies with local levels of government answerable to local electorates. Two are republics and two are constitutional monarchies. Three are unitary states while a fourth is a federation. Three have very roughly the same population while a fourth — the Netherlands — has about a quarter of that of the other three.

The **Kingdom of the Netherlands** is a unitary state divided into 11 provinces each with its own administration and a certain autonomy. The municipalities within these provinces vary considerably in size, with Amsterdam, the largest, having a population of 800,000 whereas the smallest has about 800 inhabitants. The municipalities retain a large measure of autonomy. For the purposes of water pollution control, some waters are regarded as 'state waters' and are the responsibility of a government Minister. The others are a provincial responsibility, although responsibility may be delegated to local water authorities. Only recently — and in response to EC Directives as we shall see — has the central Government been able to impose standards for water on the provinces. The provinces are also responsible for waste disposal.

The **French Republic** is a unitary state divided into 99 *départements*. Each *département* has a double administrative function. It manages departmental affairs with its own budget, with decisions being taken by the elected *Conseil Général*. In addition the *département* is a sub-division of central government. Until 1981 the Prefect, nominated by the President of the Republic, was both head of the central government administration in the *département* and also the executive responsible for carrying out the decisions of the *Conseil Général*. This double function is often difficult to understand in countries where the distinction between central and local government is so important, but practice has shown that there has not been too much confusion in France. Within the *départements* are the *communes* which can vary considerably in size and which have a locally elected *Conseil Municipal* headed by the *Maire*. The recent law on decentralisation has to some extent separated central from department government. The Prefect (renamed *Commissaire de la République*) is now only the executive of the state and occupies himself exclusively with central government affairs at the level of the *département*. Some responsibilities have been transferred to the *Maires* and some previously national functions have been transferred to the *Conseils Généraux*. The *départements* are grouped together into *Régions* which are concerned particularly with economic planning.

Control over water pollution is handled at the level of the *départements* but in addition six *Agences Financières de Bassin* have certain functions in financing investments in water pollution control. Waste is handled at the level of the commune, the *département* and the *région*.

The **Federal Republic of Germany** has a federal structure under the Basic Law of 1949 which gave a large amount of power to 11 *Laender* and indeed competence for any field rests with the *Laender* unless the Basic Law says otherwise. These 11 *Laender* have their own elected parliaments with the power to legislate and their own governments. The Basic Law divided competences between the Federation *(Bund)* and the *Laender* in three ways. In some fields (e.g. foreign affairs, defence, post and telecommunications) the *Bund* has exclusive competence. In some other fields the *Bund* and the *Laender* have 'concurrent' legislative competence. This means that *Laender* can freely exercise their competence unless the *Bund* has exercised its competence and so restricted the competence of the *Laender* in some way. By a constitutional modification of 1974 'concurrent' legislative competence was extended to the field of air pollution and the elimination of waste. Finally the *Bund* has only a 'framework-competence' in certain fields including the planning of land and the regime of water. This means that the *Bund* cannot completely regulate these fields. The *Laender* must remain free to regulate but they must respect the rules that the *Bund* has decided are essential.

The *Laender* (apart from Hamburg and Bremen) are divided into *Kreis* and the *Kreis* are divided into communes. Certain of the *Laender* are divided into three to five *Regierungs Praesident* which are purely administrative divisions and are not elected. The allocation of responsibilities for particular functions, including water and waste, between these units may vary from one *Land* to another.

The **United Kingdom of Great Britain and Northern Ireland** is a unitary state but has different administrative arrangements for Scotland, for Northern Ireland, and for England and Wales. In the past there was devolution of power to a Northern Ireland legislative assembly. There are also slightly different administrative arrangements between England and Wales, but these are not so marked as those between England and Scotland. In England and Wales there are 45 County Councils each divided into a number of District Councils. In addition the Greater London Council and the metropolitan authorities cover the big cities although these are in the process of being abolished. In Scotland there are Regional Councils and in each Region there are District or Island Councils. All these bodies are elected and have some autonomy. Waste is dealt with by counties and districts. Water pollution, however, is not handled by these local authorities but by 10 specially created river basin water authorities in England and Wales, and by river purification boards in Scotland.

# 2

# Water Pollution — Legislation and Responsibilities in the Four Countries

A large number of varied provisions are included in the many water Directives, but taken together they require the national authorities to have three basic powers if they are to be able to implement the Directives. These powers are:

- the ability to authorise discharges to water subject to conditions, i.e. to set *emission standards*
- the ability to set down water *quality objectives* for certain stretches of water for certain uses, and
- the ability to draw up *pollution reduction programmes*.

Legislation in the four countries now provides these powers but in rather different forms and with quite different administrative arrangements.

## NETHERLANDS

The present system of water quality management dates from 1970 when the Surface Waters Pollution Act 1969 was implemented. An important distinction is made in the Act between so-called 'state waters' and other watercourses. State waters comprise the largest rivers, canals and lakes together with estuaries and coastal waters. Responsibility for their management rests with the Minister of Transport and Public Works. Control of the remaining watercourses is the responsibility of the provinces though responsibility may be delegated and in eight of the eleven provinces water quality management rests with local water authorities.

The principle that all pollution- and nuisance-generating activities must be subject to prior approval was in operation well before the Directives. All discharges to surface waters require prior authorisation by the respective authority and any reasonable conditions may be attached to the licence in order to protect the environment. These conditions may include emission standards for any substances which in the opinion of the authority are likely to have a negative environmental impact. In addition, under the 1981 amendment to the Surface Waters Pollution Act, an authority is obliged to impose any relevant emission standards which are laid down by statutory instrument or ministerial decree in order to implement international agreements (including EC Directives).

For the past decade the broad policy for the control of surface water pollution has been set out in advisory Indicative Multi-Year Programmes prepared by the Minister. The first Indicative Multi-Year Programme for 1975–1979 introduced the concept of 'basic quality' — a range of parameters defining the minimum acceptable quality of all surface waters. There was, however, no means for the Minister to require the water authorities to adopt these advisory quality objectives and the 1981 Amendment Act therefore provided for a system of mandatory standards and also required the provincial and central water authorities to prepare binding water quality plans. A 1983 Decree subsequently laid down objectives for four types of surface water to be included in these plans — bathing water, freshwater to support fish life, shellfish waters and surface water to be abstracted for drinking water. Several draft plans have now been published. The plans are to determine water quality management policy for a maximum period of 10 years and are to lay down the functions allocated to surface waters, appropriate water quality objectives, priorities for the control of pollution, the general nature and scope of the necessary abatement measures and financial details.

The supply of drinking water is the responsibility of provincial water companies. Detailed provisions regarding drinking water quality are laid down in the 1984 Waterworks Decree which requires supplies to comply with a range of quality parameters and lays down monitoring frequencies and methods of analysis.

# FRANCE

The Water Law of 1964 introduced a policy of water quality objectives for each stretch of river to provide a guide for managing water. The law provided powers to fix technical specifications for water according to use, and to set time limits by which improvements were to be made. A circular of July 1971 (further elaborated in March 1978) proposed that maps of quality objectives should be drawn up and specified certain qualities for certain uses.

A Decree of February 1973 made under the Water Law requires all discharges to water to be authorised, and the law of 1976 on *Installations Classées* provides a special procedure for authorising specified plants that create grave danger or inconvenience for health, safety and the protection of nature and the environment. Discharges to water from these *Installations Classées* are authorised by reference to what is technically possible in economically acceptable conditions as determined at national level by the *Inspecteur des Installations Classées* but must also respect quality objectives.

At a national level the Ministry of the Environment is the principal coordinator of water policy exercised through the *Service de l'Eau* and the *Service de l'Environnement Industriel* which prepares the work of the *Conseil Supérieur des Installations Classées*. The Ministry of Health is responsible for drinking water.

It is at the level of the *Département* that decisions are made on authorisations and quality objectives and that standards are enforced. The *Préfet* (now called *Commissaire de la République*) — who is himself an official of the central government — has authority over all the services of the various Ministries located at departmental or regional level and relies on their advice. These include:

- Direction Departementale de l'Agriculture (DDA)
- Direction Departementale de l'Equipment (DDE)
- Direction Departementale de l'Action Sanitaire et Sociale (DDASS)
- Direction Regionale de l'Industrie et de la Recherche (DRIR)

In addition there are six *Agences Financières de Bassin* each covering one or more river basins. Their function is to finance investments for pollution reduction and to levy charges, but they do not have the power to authorise discharges. They play a co-ordinating role in drawing up maps of water quality objectives since rivers may cross several *départements*, but the responsibility for the maps and for enforcing standards rests at the level of the *département*.

The system is complicated by two factors:

- despite the *Agences de Bassin*, water quality is not controlled on a river basin basis, and there has to be discussion between *départements* on drawing up maps of quality objectives so that the process has been slow and only a small part of France is now covered by these maps
- there are two regimes for controlling discharges from *Installations Classées*. They are subjected to emission standards set nationally on the

basis of what is technically possible but should also respect quality objectives determined locally in maps of water quality. The officials responsible for quality objectives (from the DDA) are not the same as those who determine the nationally recommended emission standards (from DRIR). In practice the national emission standards are regarded as more important and often prevail over the quality objectives.

# GERMANY

Under the Basic Law water is one of the fields where the Federal Government has the right to make framework laws but where the *Laender* also retain freedom to make their own regulations subject to the framework laws. The Federal Water Law of 1976, therefore, provides a framework, and each of the 11 *Laender* has its own law which repeats and adds to the Federal law. The competence for putting the laws into effect rests with the *Laender*.

The Federal Water Law of 1976 prescribes that any use of water, irrespective of its effect on the environment, must be subject to prior authorisation. This authorisation will be given if the public interest (*Wohl der Allgemeinheit*) is not affected. For discharges there is a complementary provision: an authorisation will be given if the generally recognised technical rules are employed permitting as low a level of discharge of pollutants as possible. These rules are specified by the Federal Government in the form of minimum emission requirements. Standards now exist for 44 branches of industry and for sewage works. The responsible authorities can impose more severe standards than those laid down by the Federal Government if it is thought that the public interest is not secured by the Federal Government's minimum emission requirements.

For those industries for which minimum requirements have not been laid down the responsible authorities must themselves discover the generally recognised technical rules and must themselves fix the standards.

The Federal law also provides for water management plans. This creates the possibility of including quality objectives into these plans.

At the Federal level it is the Ministry of the Interior that is responsible for the environment and hence for water management. The Ministry of Agriculture includes fishing among its responsibilities, and the Ministry of Health is responsible for drinking water.

At the *Laender* level it is usually the Ministries with environment among their responsibilities that are responsible for water management. Each

*Land* has its own arrangement for competent authorities, but in general in the seven big *Laender* the *Kreis* share the responsibility for authorisations with the regional authorities *(Regierungs Praesident)* although the Ministry will reserve certain powers of authorisation.

A working group (LAWA) brings together representatives of the authorities responsible for water in each of the *Laender* and enables them to ensure some consistency in their practices and also to ensure some consistency in their legislation.

# UNITED KINGDOM

There are different arrangements for the administration of water in Scotland and Northern Ireland from the arrangements in England and Wales.

In 1974 10 river basin based water authorities were created in England and Wales with responsibility for all water management matters including water supply, sewage treatment, pollution control, drainage, flood prevention and fisheries.

The Control of Pollution Act 1974 is the principal Act concerned with pollution of water and it requires all discharges to inland, coastal and estuarial waters to be granted a consent, which may be given subject to conditions. This Act is only now coming fully into force and under the previous legislation discharges to coastal waters and some discharges to estuaries did not require authorisation. There are no specific powers for central government to set emission standards at national level.

The Act gives no guidance on what criteria are to be applied when authorising discharges but the water authorities have a general duty under the Water Act 1973 to restore and maintain the wholesomeness of rivers. The Water Act 1973 also places a duty on water authorities to prepare plans for water management. There are no requirements in the Act about quality objectives but in 1978 the water authorities themselves adopted a policy of publishing quality objectives for all stretches of rivers.

The Secretary of State has no specific powers to set quality objectives that water authorities must apply, but he has the power under the Water Act 1973 to give general directions to the water authorities and he can give advice in circulars.

In Northern Ireland it is the Department of the Environment for Northern Ireland that is responsible for water pollution. In Scotland it is the River Purification Boards that are responsible for controlling the quality of rivers but they are not responsible for sewage disposal or for drinking water supply.

# GENERAL COMPARISON

In all four countries there are powers to authorise discharges and in three of these, the UK being the exception, emission standards can be set at national level. Quality objectives are provided for in legislation in France and Netherlands. In Germany and UK quality objectives can be set administratively under the general powers to prepare plans for water management.

Only in the UK is water managed by authorities having complete control over a river basin. In the other three countries the authorities responsible are based on administratively rather than geographically determined areas. Although there are river basin based authorities in France which promote investments to reduce pollution, these *Agences de Bassin* do not control discharges nor themselves set quality objectives. In the Netherlands the concept of river basin management can have little meaning since the country is effectively only a small part of the basin of a great river that flows from other countries. The concept is also difficult to apply in Germany because of its federal structure coupled with the fact that powers to control water are given to the *Laender* under the Basic Law.

# 3
# Water Directives

## 3.1 DETERGENTS — DIRECTIVES 73/404, 73/405, 82/242, 82/243
### Dates for formal implementation — 27.5.75 and 8.10.83

### Brief summary of Directives

A framework Directive of 1973 prohibits the marketing and use of many types of 'hard' detergent, i.e. where the average level of biodegradability of the surfactants is less than 90 per cent. The purpose is to protect water and sewage treatment plants. One 'daughter' Directive of 1973 and two of 1982 lay down methods of testing for particular types of detergent and require 80 per cent biodegradability.

### Netherlands

#### FORMAL COMPLIANCE

The Directives are implemented by Decrees made under the Chemical Wastes Act 1976 which prohibit the marketing of certain detergents. However the Act does not extend to a prohibition on use although this technical breach of the Directive might be rectified by the forthcoming Hazardous Substances Act. Formal compliance was late.

#### EFFECT ON PRACTICE

The Directives have had no significant effect in the Netherlands since a gentleman's agreement concerned with the hardness of detergents has existed between the Minister and the producers for many years. This followed the Council of Europe's agreement on detergents of 1968.

# France

## FORMAL COMPLIANCE

The two 1973 Directives were formally implemented by Decrees of 1977. The 1982 Directives are providing the opportunity to revise the Decrees and to replace them with a single Decree. Formal compliance was late.

## EFFECT ON PRACTICE

The Directives have resulted in changes to the composition of detergents sold except in the case of non-ionic detergents. Measurements of river quality show a decline in detergents since the introduction of the Decrees resulting from the Directives which suggests that they are having a beneficial effect.

# Germany

## FORMAL COMPLIANCE

With its law on detergents of 1961 the FRG was the first Member State to legislate for biodegradability, but this had to be amended in 1975 to conform with the 1973 Directives. A decree of 1977 set a level of biodegradability of 80 per cent. Formal compliance was therefore late.

## EFFECT ON PRACTICE

The great improvement in biodegradability followed the 1961 law and the Directives have not had much effect on practice.

# United Kingdom

## FORMAL COMPLIANCE

Regulations were made in 1979 to implement the two 1973 Directives and in 1984 to implement the two 1982 Directives so formal compliance was late.

**EFFECT ON PRACTICE**

A voluntary agreement between government and the producers had existed at least since 1964 and the standards of the Directives were met before the Directives were agreed. They have thus had no effect on the quality of detergents manufactured in Britain, but can be used to control imports.

## General assessment

The Directives have applied a stimulus to improve the biodegradability of detergents in France, but not in the other three countries where legislation (FRG) or voluntary agreements (NL, UK) already existed. The Directives will have ensured a more uniform system throughout the Community and can be used to control imports which are not easily controlled by voluntary agreements. Formal compliance with the Directives was late in all four countries. Some confusion over formal compliance has been caused by the requirement of the framework Directive that a biodegradability of 90 per cent should be achieved whereas the daughter Directives specify 80 per cent.

## 3.2 SURFACE WATER FOR DRINKING — DIRECTIVE 75/440 Date for formal implementation — 18.6.77

## Brief summary of the Directive

Sources of surface water for the abstraction of drinking water are to be classified into three categories — A1, A2, A3 — corresponding to the three standard methods of treatment for drinking water. Forty six 'parameters' are listed with 'imperative' and 'guide' values for each category A1, A2, A3. Member States must set values defining the water quality at abstraction points no less stringent than the 'imperative' values. Ninety five per cent of samples must meet these values. The use of surface water worse than A3 is prohibited except in exceptional circumstances and the Commission must be informed. A plan of action with a timetable for improvement, especially for A3 water, must be prepared (although the obligation is qualified).

# Netherlands

## FORMAL COMPLIANCE

The Surface Waters Pollution Act 1969 had to be amended to empower central Government to set water quality objectives. This did not happen until 1983 after the Netherlands had been found to be in breach of the Directive by the European Court. The Court rejected the argument that the 1969 Act provided sufficient powers and that the Indicative Multi-Year Programme for water management gave effect to the various quality objectives since the Government had no powers to oblige the water authorities to adopt the necessary quality objectives. The new legislation sets January 1985 as the date by which surface water standards broadly corresponding to A2 quality are to be met. This new legislation is broader than the Directive in that it applies to surface water which is abstracted after percolation through the ground or through artificial filters. The Directive does not regulate this source of drinking water (but see comments on Germany below). Improvement plans are required by the 1981 amendments to the Surface Waters Pollution Act. Amendments made in 1984 to the Waterworks Decree prohibit the preparation of drinking water from surface water worse than A3.

## EFFECT ON PRACTICE

If it had not been for Community Directives Dutch water management policy would have been based on the concept of 'basic quality' — a minimum standard applied to all surface water in the country. Therefore the separate quality standards for surface water intended for drinking can be attributed entirely to the Directive. However the Directive is not as significant as this might imply since wherever possible Dutch water companies (see also Germany below) use groundwater or water percolated through the ground which is outside the scope of the Directive. Some upgrading will be necessary since some abstraction points are worse than A3 quality. In only one of the regional plans which were published in 1984 was surface water abstracted for drinking and designated as such. The Public Works Department designated six waters in its plan for state waters.

# France

## FORMAL COMPLIANCE

There is no French legislation concerned specifically with the quality of surface water for drinking but merely an obligation under the *Code de la Santé Publique* that drinking water must be *potable*. Water quality objectives are provided for in the Water Law of 1964 and elaborated in a 1971 circular. A further circular of 1978 required water quality maps for each *département* specifying A1, A2, and A3 qualities. There was a delay in formal compliance since the date set in the Directive is July 1977. The purpose of these maps is to guide investments in sewage works and authorisations for discharges to water and abstraction. Discharges are controlled under the 1964 law on water. These maps effectively constitute programmes since they include timetables for improvement.

## EFFECT ON PRACTICE

The idea of water quality objectives existed in France well before the Directive but was refined in a circular in 1978. Although the circular does not mention the Directive it adopts the same standards and was influenced by it. The preparation of water quality maps is far from complete since agreement must be reached between a number of *départements*. There is no requirement in the Directive that Member States should send a list of abstraction points to the Commission listing their quality (A1, A2, or A3) but a list has been sent to the Commission. The Directive has obliged suppliers of water to use treatment methods appropriate to the source, and they have had to abandon certain sources worse than A3. The Directive has had a significant influence in France and constitutes a point of reference for those managing water.

# Germany

## FORMAL COMPLIANCE

There is no specific legislation on the quality of surface water for drinking. The federal water law of 1976 requires all discharges to be authorised and requires management plans to be made. A federal regulation of 1978 sets out the minimum content of management plans and sets out seven classes of water quality. The *Laender* in turn have sent their own circulars to the competent authorities which vary slightly but which describe the Directive. Some say that waters worse than A3 may not be used for drinking water, and that values must be fixed for abstraction points. The values are fixed in abstraction authorisations.

In Germany water is often taken from rivers and is then injected into groundwater before being abstracted so as to benefit from filtration. Such water is excluded from the scope of the Directive. Some water is also abstracted from the banks of rivers so as to benefit from filtration. There would be ambiguity about whether this falls within the Directive, but some of the *Laender* circulars explain that a statement in the minutes of the Council meeting (which are otherwise confidential) says that the Directive is to be regarded as a Recommendation for water filtered by river banks. Other *Laender* circulars say that the Directive does not apply to those waters.

## EFFECT ON PRACTICE

In Germany 71 per cent of drinking water comes from groundwater and 29 per cent from surface water (1 per cent running water, 3 per cent lakes, 7 per cent reservoirs, 6 per cent filtered through banks, 12 per cent injected into groundwater). Thus only 11 per cent is subject to the Directive. Since the water in reservoirs is already of a higher standard in effect only four per cent of drinking water (some say one per cent) is influenced by the Directive. In some *Laender* no waters are subject to the Directive. Most sources are of A1 or A2 quality, very few of A3 (e.g. Lake Constance). No list of sources has been sent to the Commission. It seems that only one derogation has been granted by the Commission for a transitional period until another better source became available. There are no improvement plans for drinking water as such, but for Lake Constance where the quality makes such a plan obligatory there is a general improvement plan involving neighbouring countries too.

The limited amount of water subject to the Directive means that it can have no great effect in improving surface water quality. However the values in the Directive are frequently used as a reference for purposes other than those that are obligatory under the Directive. The Directive with other water quality Directives undoubtedly influenced the law giving the *Laender* the powers to prepare water management plans.

# United Kingdom

## FORMAL COMPLIANCE

Water authorities have powers to authorise discharges to water and powers (under the Water Act 1973) to produce improvement plans. They do not have specific powers or duties to set water quality objectives nor has the Secretary of State. However, water authorities can set water quality objectives as part of their improvement plans. The Directive was accordingly complied with by the Department of the Environment informing the water authorities of the Directive and advising them of the steps to take. The Secretary of State can if necessary give general directions to the water authorities. Separate action was taken in Scotland and Northern Ireland.

## EFFECT ON PRACTICE

About 70 per cent of drinking water in Britain comes from surface water. A list of over 700 abstraction points has been sent to the Commission. Of these only 10 are A3 and none are worse than A3. All authorities have set imperative values equal to those in the Directive, some have set guide values more stringent than those in the Directive. The improvement plans have been integrated with the forward plans that authorities produce anyway. In a very few cases the treatment given to water was below that required by the Directive which has therefore drawn attention to the need for some improvements. Many improvements were however programmed anyway.

# General assessment

There is an ambiguity about the Directive. Is it intended to secure the quality of drinking water, or is it intended as a stimulus to improve the quality of surface water? If it is intended to improve rivers it has probably not been very successful since so many abstraction points fall outside its scope. This is because the definition of surface water excludes river water filtered through river banks or injected into groundwater, a fact which is not completely clear from the text of the Directive itself. No complete list of abstraction points has been sent to the Commission by all Member States to enable a comparison to be made. Although such lists are not required by the Directive they would be a valuable indication of how unevenly the Directive bears on different Member States as well as providing information about the sources of drinking water in the Community. The Commission has an obligation to compare the various improvement plans but has not yet done so.

The major effect of the Directive (taken together with other water Directives) has probably been to promote discussion of water quality objectives for different uses in the Member States. Only in France was this idea formalised in legislation before the Directive, but even in France it was not well developed so that the Directive has had a significant influence. Nor was the idea of quality objectives well developed in the other three countries. Although in the Netherlands a 'basic quality' for surface water had been laid down it was not related to specific uses. In the Netherlands national legislation has had to be introduced and because of the lateness of the Dutch legislation empowering the Government to set national water quality objectives the Commission took the Netherlands' Government to the European Court. The difference between the position in the Netherlands and the UK is that although the UK government has no powers to set national quality objectives it nevertheless has the power to compel the water authorities to comply with objectives that can be set administratively. This was not so in the Netherlands before the new law.

## 3.3 SAMPLING SURFACE WATER FOR DRINKING — DIRECTIVE 79/869
Date for formal implementation — 10.10.81

## Brief summary of the Directive

The Directive supplements the surface water Directive (see 3.2) by recommending methods of measuring the parameters for surface water quality (i.e. the methods are not obligatory), and by setting the frequencies for such measurements.

## Netherlands

**FORMAL COMPLIANCE**

The Quality Objectives Decree of 1983 prescribes the methods of measurement and the frequencies in terms that comply with the Directive. Formal compliance was thus late.

**EFFECT ON PRACTICE**

The Directive is leading to a single monitoring regime being applied by all authorities. Previously there were differences in the parameters sampled and methods of analysis applied.

## France

**FORMAL COMPLIANCE**

A proposed Decree and an *arrêté* were in course of preparation at the end of 1984 to implement the Directive. Formal compliance was therefore late.

**EFFECT ON PRACTICE**

Until the measures are in place it is not possible to judge their effect although some effect is likely since not all the parameters were previously measured.

# Germany

## FORMAL COMPLIANCE

Formal compliance caused no problem and was achieved within the prescribed period. The federal water law of 1976 and the water laws of the *Laender* provide the necessary powers. The *Laender* working group on water (LAWA) drafted an implementing text that all the *Laender* used in their circulars to the competent authorities.

## EFFECT ON PRACTICE

The reference methods of the Directive are not used but instead equivalent German methods. Not all the parameters were previously measured. The methods of the Directive and many parameters are much criticised. The Directive has therefore had very little effect. Suggestions have been made by the Federal Government that the Directive should be replaced by a sampling Directive applicable to all the water Directives.

# United Kingdom

## FORMAL COMPLIANCE

Legal powers to monitor water quality already existed. The formal step taken to comply with the Directive was a circular from the Department of Environment sent to the water authorities explaining the Directive and appointing them as competent authorities. A similar circular was issued in Scotland.

## EFFECT ON PRACTICE

The Directive has had little effect in Britain. The frequencies of measuring commonly applied in Britain are much greater than set out in the Directive. The greatest effect is that in some cases extra parameters have had to be measured which were not previously measured.

# General assessment

As originally proposed the Directive would have made obligatory certain measuring methods but this was much criticised (e.g. in FRG and UK) and as a result the Directive is in effect little more than a recommendation since the measuring methods do not have to be used and only the frequencies are obligatory. The Netherlands was late in complying with the Directive and in France the Directive has still not been formally complied with. In probably all the four countries it has resulted in some more parameters being measured since not all were measured before.

## 3.4 DRINKING WATER — DIRECTIVE 80/778 Date for formal implementation — 17.7.82

## Brief summary of Directive

Member States are to ensure that by July 1985 drinking water meets certain values for over 60 parameters listed in an Annex. The values are to be fixed by the Member States and must be as stringent as certain 'Maximum Admissible Concentrations' (MAC) figures. More stringent 'Guide Levels' (GL) which Member States may adopt are also listed.

Member States may *derogate* in two circumstances: related to the nature of the ground (permanent derogations), and related to situations arising from exceptional meteorological conditions (temporary derogations). The Commission is to be informed of derogations. In emergencies the MAC figures may be exceeded for limited periods if there is no risk to health and water cannot be supplied in any other way. They may also be exceeded subject to conditions where a source worse than A3 has to be used (see Surface Water Directive — 3.2).

Member States may also ask the Commission for a *delay* in meeting the MAC values in exceptional cases but must propose an improvement programme with a timetable. Requirements for monitoring and analysis are set out in two further Annexes.

## Netherlands

### FORMAL COMPLIANCE

The Waterworks Act 1957 regulated the management of drinking water supplies and the Waterworks Decree 1960 the quality of drinking water. An amended Decree came into effect in July 1984, i.e. two years after the date required by the Directive and after the Commission had started infringement proceedings against the Netherlands. The new Decree closely follows the Directive.

## EFFECT ON PRACTICE

The 1960 Waterworks Decree laid down MAC values for seven substances of which only those for nitrites, cyanide and chromium were equal to or tighter than those in the Directive. In practice, however, the management of drinking water supplies has long been guided by the broader recommendations of the Association of Water Company Proprietors. These included 14 parameters, the majority of which are comparable to or tighter than those in the Directive. With over 60 parameters the Directive would appear to make a significant change but the Association of Water Company Proprietors concludes that in practice the Directive will not lead to significant improvement. Some difficulties will nevertheless arise, most notably with nitrates. The nitrate level is exceeded at some groundwater pumping stations but no derogations or requests for delays are being made since the standards are to be achieved by blending different sources.

# France

## FORMAL COMPLIANCE

There is a general obligation in the *Code de la Santé Publique* that drinking water should be fit for human consumption, and a Decree of 1961 makes further provisions. This 1961 Decree is to be replaced by another specifying the MAC values of the Directive. Formal compliance is therefore late. The new Decree is presently being considered by the *Conseil Supérieur d'Hygiène*, and it is likely to say that, although new water supplies may not be authorised unless the nitrate value is less than 50 mg/l (the MAC in the Directive), existing supplies with a value between 50 mg/l and 100 mg/l may continue to be used. This will conflict with the Directive and a delay will have to be sanctioned by the Commission.

## EFFECT ON PRACTICE

A great deal of attention has been focussed on the problem of nitrates as a result of the Directive since two per cent of the population receive water with a nitrate level between 50 mg/l and 100 mg/l and some with a level over 100 mg/l. In addition there are likely to be problems with organo-chlorine compounds, aluminium and bacteriological parameters. There are plans to make derogations for certain parameters and to request the Commission for delays. The monitoring and analysis requirements of the Directive will not pose problems. The main impact of the Directive on the quality of water depends on the finance for treatment plants, a matter which has been given greater attention at national level as a result of the Directive.

# Germany

### FORMAL COMPLIANCE

German law does not contain general provisions on the quality of drinking water. The relevant provisions are to be found in a number of different texts. There is a decree on the treatment of drinking water which lists substances that can be used for treatment and the maximum that may remain. The law on epidemics specifies that water must not put public health at risk. The decree on drinking waters sets parameters that must be analysed — 11 substances are given maximum values — the value for nitrates being 90 mg/l.

German law does not therefore comply with the Directive. A new draft decree is being considered now by professional bodies and the *Laender*.

### EFFECT ON PRACTICE

The biggest practical problem posed by the Directive is the need to respect certain values particularly the nitrate value. Even the current value for nitrates (90 mg/l) is sometimes exceeded. According to the *Laender* between 1% and 3.5% of drinking water distributed through public supplies exceeds the level of 90 mg/l and a higher proportion of private supplies. There will be problems with meeting certain other values in the Directive, e.g. chlorinated hydrocarbons and iron and manganese. The full impact of the Directive will not be known until the new decree is adopted and begins to take effect.

# United Kingdom

### FORMAL COMPLIANCE

The Water Acts of 1945 and 1973 place a duty on water authorities to supply 'wholesome' drinking water and a duty on the local authority to ensure that it is 'wholesome'. Any dispute between water authorities and local authorities is determined by the Secretary of State. There are no mandatory standards in British legislation defining what is to be regarded as 'wholesome' although, before the Directive, the World Health

Organisation's European Standards were generally used as guidelines. In August 1982 the Department of the Environment issued a circular saying that no new legislation was necessary and that the Secretary of State would regard 'compliance with the terms of the Directive as a necessary characteristic but not a complete definition of any water that is to be considered wholesome', and that this is how he would interpret 'wholesome' in any dispute referred to him. The circular explains the Directive and that the standards must be met. Since the Secretary of State has power to compel a water authority to take certain action, it is the view of the Government that a circular is sufficient for formal compliance. Further circulars have been issued elaborating on certain points, for example, fixing the maximum values where the Directive only specifies a Guide Level (GL).

## EFFECT ON PRACTICE

The circular of 1982 said that the Directive would 'underline and reinforce, rather than alter, existing policy and procedure'. The Directive nevertheless creates some additional pressures. Thus a four year delay is being sought from the Commission for meeting the lead parameter accompanied by a national programme for improvement, and a delay of at least 10 years is being sought for most private water supplies serving under 500 people. Delays or derogations are also likely to be sought for individual public supplies containing substances above the level in the Directive. This will be so in the case of nitrates for several public supplies.

# General assessment

In France and Germany formal compliance is not yet complete and in the Netherlands it was late. New legislation (a Decree) had to be introduced in the Netherlands and new legislation is necessary in both Germany and France. The UK has relied on administrative measures, i.e. circulars giving advice on applying existing broad legislation.

The date by which the standards are to be met is July 1985 and any derogations have to be communicated to the Commission and any delays requested. Delays have to be accompanied by improvement programmes. It is already clear that delays or derogations will be required in several countries for some parameters particularly the nitrate parameter.

The Directive is having important effects in all four countries although its full impact will be easier to assess once the extent of delays and derogations are known. A particular effect has been to stimulate discussions in all countries on the increase in nitrate levels and the validity of the 50 mg/l standard is being called into question.

## 3.5  WATER STANDARDS FOR FRESHWATER FISH — DIRECTIVE 78/659
Date for formal implementation — 20.7.80

## Brief summary of Directive

Member States are to designate freshwaters needing protection or improvement in order to support freshwater fish in two categories: suitable for salmonids and for cyprinids. Values are to be set for 14 parameters at least as stringent as the I (imperative) values in an Annex although more stringent G (guide) values are to be respected if possible. Pollution reduction programmes are to be established to ensure that within five years the waters conform to the values. Minimum sampling frequencies and reference methods of analysis are set. Derogations may be given for certain parameters in certain circumstances. Member States are to send the Commission a list of designated waters and five years later a detailed report.

## Netherlands

### FORMAL COMPLIANCE

As with the surface water Directive (see 3.2) no mechanism existed for setting and enforcing national standards until recently. These were made possible by the 1981 amendments to the Surface Waters Pollution Act. In 1983 the Quality Objectives Decree laid down standards which are to be applied to freshwater designated for fish life. These standards are to be laid down in the various water quality plans, and water must comply within five years. The delay in formally implementing the Directive thus exceeded three years. The I values laid down are generally tighter than in the Directive. The sampling frequencies and methods are similar to the Directive except in two respects.

### EFFECT ON PRACTICE

Together with the other Directives on water quality objectives (3.2, 3.6, 3.7) this Directive has been responsible for the shift towards the functional approach in Dutch water management. This change in policy may only have a limited practical effect with fish waters because in only five cases do the values for cyprinid water significantly improve upon or add to the requirements of the standard 'basic quality' criterion which previously existed. No list of designations has yet been sent to the Commission. The full impact of the Directive can be judged when designations have been made, and it is not clear from the draft water quality plans presently available to what extent designations are to be made. It seems unlikely that any waters for salmonids will be designated and the intention seems to be to designate only those waters which already closely comply with the given values.

# France

## FORMAL COMPLIANCE

The 1964 water law provides for quality objectives and these were developed in a circular of March 1978 which set out the procedure for preparing maps of quality objectives. A circular of December 1978 referred specifically to the Directive and said that designations were to be made on the basis of information and proposals from the *services de polices des eaux* and conforming with the March 1978 circular on the policy of quality objectives. The December 1978 circular said that the I values of the Directive would serve to define the objectives and that the G values could be used depending on local circumstances. An *arrêté* of May 1983 formalised the status of the departemental water quality maps, and made clear that they represent a forward programme. It also explained the procedure for consultation before the maps are officially approved. Although some Decrees have been made establishing quality maps and improvement plans for certain rivers — some of which refer to the Directive — there is no list of designations for France in the terms set out in the Directive. A circular of November 1984 refers to the Directive and requires a list of designated stretches to be sent to the Ministry of the Environment by the *départements* by the end of 1985.

## EFFECT ON PRACTICE

The procedure for drawing up water quality maps is complicated and requires consultation between different bodies. Two further explanations can be put forward for the failure to designate fresh waters for fish. There is resistance from industry which is concerned at the costs that designation may entail, and there is resistance from fishermen who believe that stretches of water not designated will be sacrificed. Although some stretches of water have been 'designated' for fish these are regarded as 'technical' designations and not subject to the provisions of the Directive.

There are also many criticisms of the Directive based on technical grounds. Action is now being taken by the *départements* following the circular of November 1984 (see above).

# Germany

## FORMAL COMPLIANCE

The 1976 law on the management of water provides for the establishment of plans for the management of water and requires discharges into water to be controlled. The legal powers to put the Directive into effect therefore exist but no circulars or other administrative documents specific to the Directive have been issued by the *Laender*. No waters have been designated. Since it is within the discretion of Member states not to designate any waters it is argued that formally the Directive is complied with at the Federal level.

The failure to designate could have been anticipated from the parliamentary discussions in Germany that preceded adoption of the Directive. The *Laender* were firmly opposed on the grounds that the protection of a single function is not a good approach to the protection and improvement of the quality of water. The Federal Government appears only to have agreed to the Directive to demonstrate a good Community spirit.

## EFFECT ON PRACTICE

There have been no effects on practice except that the values in the Directive are sometimes used as a reference.

# United Kingdom

## FORMAL COMPLIANCE

Powers to control discharges to fresh water and powers to make plans to improve water already existed. In October 1978 the Department of the Environment sent a copy of the Directive with a circular letter explaining it and stating that the functions involved were being delegated to the water authorities. The circular advised water authorities to consult local authorities and fishing and other environmental interests on the designation of water and on other matters. A similar letter was issued in Scotland.

**EFFECT ON PRACTICE**

The circular letter of October 1978 suggested the setting up of a working group with representatives of government departments and water authorities to prepare more detailed advice on the Directive. The resulting advice note stated that the aim should be to designate as many waters as possible without increasing expenditure, in other words to designate only waters that already met the standards or would do so by 1985 under existing plans. This has happened. The list of designated waters sent to the Commission in 1980 included very few that did not already meet the standards. Nevertheless designations are extensive and some water authorities have designated 50 per cent of their total length of rivers. Some water authorities issued consultation papers setting out their intended designations and made changes as a result of representations received from fishing interests. Some additional designations were made after 1980. A report is now being prepared — probably in map form — to be sent to the Commission by July 1985. Some authorities have had to monitor for substances they did not monitor before.

The Directive has probably not yet had any effect on water quality, but it does mean that it is more difficult for the designated waters to be allowed to deteriorate. It also means that the Department of the Environment and the water authorities have a much better understanding of the extent of good waters for fish and they also now have a standard for making comparisons.

# General assessment

The Directive has not yet had any effect in improving water quality. In France and Germany no waters have so far been designated. In the Netherlands designations are only just being made. In the UK extensive designations were made in 1980 but nearly all met the standards and the rest were programmed for improvement anyway. Its main benefit is therefore to prevent the designated waters being allowed to deteriorate and to increase understanding of the quality of British waters.

The delay in formally complying with the Directive in the Netherlands can be explained by the need to introduce new legislation to adapt Dutch practice to the new functional approach required by this and the other Directives, i.e. the setting of water quality objectives according to use.

The failure to make designations in France can be attributed to the difficult procedure required in drawing up the water quality maps and the opposition of both industrialists and fishermen.

In Germany there was opposition to the Directive from the outset among the *Laender* which they have maintained by making no designations. The position in Germany results from its federal structure. It is the Federal Government that negotiates and agrees to Directives and therefore may be assumed to have some commitment to implementing them both formally and in practice, but it is the *Laender* who have competence for water matters. It is possible that when they see that other countries have designated waters without great difficulties they may review their opposition. The Directive even if implemented by just designating waters already conforming to the standards would at least ensure a better understanding of the quality of waters in Europe.

The risk that some Member States would not make designations and so render it inoperative was foreseen. This problem could have been overcome by appropriate draughtsmanship. For example the Directive could have required the Member States to make at least some designations (as with the bathing water Directive) while still giving them a considerable degree of latitude.

A further comment can be made about this Directive. The British have argued in favour of the quality objective approach and therefore have been motivated to prove their commitment. The Germans on the contrary have not believed in the quality objective approach and have effectively ignored the Directive. The French who first had the quality objective approach in their legislation have been hampered by the complexities of their system for water administration from putting it into effect easily. The Dutch have had to change their approach and to introduce new legislation which has entailed delays.

# 3.6 SHELLFISH WATERS — DIRECTIVE 79/923 Date for formal implementation — 5.11.81

## Brief summary of Directive

This Directive is similar in form to the freshwater fish Directive (3.5). The Member States are themselves to designate coastal and brackish waters which need protection or improvement so as to support shellfish. Initial designations were to be made by November 1981 but additional designations can be made subsequently. Member States are to establish pollution reduction programmes so that within six years of designation the waters conform to certain values set for 12 parameters. The Directive sometimes specifies I (imperative) values and sometimes G (guide) values for the parameters. (There is only a G value for faecal coliforms and not an

I value.) Frequencies of sampling are specified. A list of designations is to be sent to the Commission followed, six years later, by a report on the designated waters.

# Netherlands

## FORMAL COMPLIANCE

The 1983 Quality Objectives Decree made under the 1981 amendments to the Surface Waters Pollution Act laid down the values for the parameters for shellfish waters that are to be adopted by the relevant implementing authority when designating shellfish waters. In this case the Minister himself is the only water authority concerned since coastal waters alone are involved. Formal compliance was thus two years later, and no initial designations were made by November 1981.

## EFFECT ON PRACTICE

It is not clear from the Public Works Department's draft water quality plan for state waters exactly which shellfish waters are to be designated, and the effect of the Directive cannot therefore be properly assessed. There are nevertheless indications that large stretches of coastal waters with mussel beds will be designated including the whole of the Wadden Sea. However, these waters already largely comply with the necessary quality objectives. Any improvement in Dutch coastal waters is also largely outside Dutch hands because of the influence of the Rhine.

# France

## FORMAL COMPLIANCE

No shellfish waters have yet been formally designated in accordance with the Directive and no legislation has been specifically introduced to implement the Directive. An interministerial circular of May 1982 with title *'Application de la Directive Européenne relative a la qualité des eaux conchylicoles'* proposed the formation of a committee under the authority of the Prefect of the *département* to give advice on the designation of zones proposed by the *Institut Scientifique et Technique des Pêches Maritimes*. These proposals are then to be submitted for an opinion to the *Conseil Général* of the relevant *départments* who then approve the designations. If the *Conseil Général* does not do so, in the last resort the Minister of the Environment has the task of arbitration. There is legislation that predates the Directive providing powers to implement the Directive.

## EFFECT ON PRACTICE

The Directive has provided a stimulus to a study of areas for designation. Criteria have been established which have enabled some areas to be eliminated, either because there is little exploitation of shellfish or because of the poor quality of the water. So far four *départements* have designated shellfish waters which are to be the subject of an *arrêté ministeriel* whereupon they will be formally communicated to the Commission. For all coastal *départements* pollution reduction programmes or monitoring programmes have been established or are to be. For example in Finistere (Brittany) pollution reduction programmes costing 200 million francs have been proposed for four principal shellfish waters. Cost at a time of financial restraint is a major problem.

# Germany

## FORMAL COMPLIANCE

No legislation or administrative measures have been adopted by the Federal Government or by the *Laender* to implement the Directive, and no designations have been made. The Directive only affects the four coastal *Laender* and two of these (Hamburg and Bremen) have informed the Federal Government that they have no shellfish waters. The Federal Government has said to the Commission that the 1976 law on the management of water and the laws of the *Laender* of Lower Saxony and Schleswig Holstein contain the provisions necessary for formally implementing the Directive, since they provide for the establishment of plans for the management of water and require discharges into water to be controlled.

## EFFECT ON PRACTICE

The *Laender* were opposed to the Directive when it was proposed, and have maintained their opposition. The main objection now made is that quality objectives formulated by reference to the use of water or by reference to the life of a single aquatic species are not a good approach to the protection of water.

The only two *Laender* with shellfish waters (Lower Saxony and Schleswig Holstein) have not designated shellfish waters and have no plans to do so.

However both have programmes as a result of the Paris and Oslo Conventions and in the framework of a programme for the North Sea, and claim that these are more stringent than the requirements of the Directive and satisfy its monitoring obligations. Nevertheless the Directive provides a certain pressure.

# United Kingdom

## FORMAL COMPLIANCE

Under the Control of Pollution Act all discharges to estuaries and coastal waters have to be authorised. However the Act is only now coming fully into force and under the previous legislation discharges to coastal waters did not have to be authorised, nor did all old discharges to estuaries. So long as designated shellfish waters were not polluted by a discharge not controlled under the previous legislation it can be held that the legislation is sufficient for formal compliance. In January 1980 the Department of the Environment sent a copy of the Directive to the water authorities with a circular explaining it and saying that they would be responsible for implementing it. A similar letter was issued in Scotland.

## EFFECT ON PRACTICE

The circular of January 1980 proposed a working group with representatives of government departments and water authorities to prepare detailed advice on the Directive. Similar action was taken in Scotland. The resulting advice was that the initial designations should not put an extra burden on capital expenditure given the economic circumstances and that only a fairly small number of waters should be designated. These were to be those which either already met the standards or which were capable of doing so by October 1987 under existing plans. This advice was followed and a list of 27 designated shellfish waters (totalling 314 square kilometres) in England and Scotland was sent to the Commission in November 1981. In January 1983 an additional designation in Northern Ireland and another in Wales were communicated to the Commission. Some additional expenditure has been incurred in sampling and monitoring although its greater effect has been to redistribute existing monitoring efforts. Preparatory work has begun for the detailed report on designated waters that has to be sent to the Commission in 1987.

The Directive has also had the effect of applying pressure for the full implementation of the Control of Pollution Act and has forced water authorities to involve themselves with sea water earlier than they would otherwise have done.

# General assessment

The Directive has not yet had any effect in improving water quality. In France, Germany and Netherlands no waters have so far been formally designated and the designations in the United Kingdom are all for waters that already met the standards or were programmed for improvement anyway. In France and Netherlands the process of designating waters has started but there are no plans to designate waters in Germany.

The reasons for failure to designate water in three of the countries are similar to those applicable to the freshwater fish Directive (3.5). The Netherlands had to introduce new legislation which has entailed delays. In France the procedures are complicated, since a large number of authorities and interests are involved, and have resulted in delay. In Germany the relevant *Laender* do not believe that quality objectives formulated by reference to the life of a single aquatic species are appropriate for protecting water and do not intend to make any designations.

Britain which has argued in favour of quality objectives has been motivated to prove its commitment to the idea by making designations within the prescribed time, but the designations made have been limited in extent. The greatest effect in Britain has been to force the water authorities to pay greater attention to seawater before being required to do so by the Control of Pollution Act which is only now coming fully into effect.

# 3.7  BATHING WATER — DIRECTIVE 76/160
## Date for formal implementation — 10.12.77

## Brief summary of the Directive

Bathing water is defined as fresh or sea water (excluding swimming pools) in which:

- bathing is explicitly authorised, or
- is not prohibited and is traditionally practised by a large number of bathers.

Member States must set values for a number of parameters (including coliforms) against which an I (imperative) and G (guide) value is given. The values must be met by December 1985 but derogations may be granted if accompanied by a management plan. Sampling frequencies and methods are specified but no conditions are laid down for handling samples before analysis Regular reports are to be submitted to the Commission.

# Netherlands

## FORMAL COMPLIANCE

Two separate, but overlapping, bodies of legislation regulate bathing waters: that relating to surface water management and that relating to bathing facilities. The 1982 Water Quality Plans Decree made under the 1981 amendments to the Surface Waters Pollution Act requires water authorities to allocate functions to surface waters including bathing where bathing is or will be practised by a considerable number of people. The 1983 Quality Objectives Decree lays down quality objectives in general following the values of the Directive except for coliforms and faecal streptococci. The explanatory memorandum says that for these the values of the Directive are impossibly stringent. The delay of five years resulted in a case being brought by the Commission before the European Court for non-compliance. The Court rejected the argument of the Netherlands that compliance had been secured by incorporating the quality objectives in the Indicative Multi-Year Programme on the grounds that the Programme was not binding.

For many years the legislation relating to the hygiene and safety of swimming pools has also applied to the many hundreds of specially adapted freshwater bathing areas — often operated commerically. The 1982 Hygiene and Safety of Bathing Facilities Act now extends to areas not specially adapted but where bathing is practised by a considerable number of people. The Hygiene and Safety of Bathing Facilities Decree of November 1984 sets out the same quality objectives for both adapted and the other bathing waters which are to be met by 1986, i.e. in accordance with the Directive. The limits are similar to the I values in the Directive. The Act requires provinces to draw up a list of waters specially adapted for bathing and those where bathing is practised by a considerable number of people.

The values in the Quality Objectives Decree are in general stricter than those in the proposed Bathing Facilities Decree and the explanatory memorandum accompanying the Bathing Facilities Decree discusses the overlap.

**EFFECT ON PRACTICE**

Seven of the nine draft water quality plans published in 1984 set out a list of designated bathing waters. These seven provinces are to adopt the values of the Quality Objectives Decree rather than the less strict values of the proposed Bathing Facilities Decree. It is proposed that the whole of the Dutch coastline will be designated as a bathing water in the water quality plan for state waters.

From interviews it appears that a variety of interpretations are being given to the term 'considerable number'. The number of bathers to be found at a single site on a typical summer's day is the decisive factor in designating the area, but the minimum number adopted by water authorities varies between 10 and 100 and is sometimes vaguer. However, it is also apparent that the existing quality of the water is an important consideration in practice since many poor quality waters which are presently used by large numbers of bathers will not be identified as bathing waters because the costs of improvement are too high. For these waters bathing will have to be prohibited. Thus the Directive makes an important shift in Dutch practice: until now bathing has taken place in areas where it is explicitly authorised and in areas where it has been traditionally practised. Under the new arrangements bathing may only be practised in waters explicitly designated for the purpose in a water quality plan; bathing is no longer to be tolerated where it is not so authorised. The Directive is therefore likely to reduce the number of bathing areas rather than to improve the quality of bathing water. Coastal waters are an exception to this trend although the task of bringing water quality along the entire coastline up to standard is considerable. No derogations were applied for.

# France

**FORMAL COMPLIANCE**

Before the Directive fresh bathing waters were controlled under an *arrêté* of June 1969 and sea water by reference to recommendations of the World Health Organisation. The 1969 *arrêté* was replaced by one of April 1981 made under the *Code de la Sante Publique* which defines the bathing water quality standards of the Directive. The *arrêté* also requires existing discharges affecting a bathing water to be improved if the standards are not met, and new discharges not to be authorised unless they do not affect the quality of a bathing water. Formal compliance was accordingly late by over three years. The *arrêté* also requires an inventory of bathing waters actually used in contrast with the list of authorised bathing waters that previously existed. The inventory and sampling is undertaken by the *Directions Départmentales des Affaires Sanitaires et Sociales*.

## EFFECT ON PRACTICE

During 1981 3,209 sampling points divided amongst 1,823 communes were checked and 26,493 samples taken (an average of 8.25 per sampling point). No national criteria have been laid down for 'large numbers of bathers' and the criteria used by the *départements* and *mairies* for identifying bathing waters are:

- accessibility to the public and a significant use by bathers,
- the willingness of the municipality to manage and supervise.

In 1981 about 25 per cent of the sampling points did not conform to the standards of the Directive. The sampling results are in general made public by notices where bathing takes place, in the *mairies* and in the press. Where pollution is established the *maire* must prohibit bathing. A national summary is prepared at the end of the season and is sent to the Commission. The measures taken or proposed to meet the standards in the Directive include improvements to sewage systems and the connection of households to sewers so that they do not discharge direct to sea. In the Alpes Maritimes, for example, four major sewage works were started in 1982–3 and three more are planned for 1985. The Directive has undoubtedly ensured that certain schemes have been proceeded with more quickly than would otherwise have been the case.

For fresh waters there are in place considerable difficulties in meeting the standards such as in the Seine at the confluence with the Orge.

No derogations were communicated to the Commission in 1981.

# Germany

## FORMAL COMPLIANCE

Before the Directive bathing waters were sampled but there were only recommended microbiological standards. However the Federal Law and those of the *Laender* are sufficient for formally implementing the Directive, particularly the power for management plans. All that was required was administrative measures in the *Laender*, but these were not all ready by 1977. The final measures were not communicated to the Commission until 1980 so that formal compliance was late. These measures are generally similar. They fix the values of the Directive, require the identification of bathing waters, and appoint the competent authorities for identification and for control.

## EFFECT ON PRACTICE

All the *Laender* have identified bathing areas following criteria proposed by LAWA (a joint working group of the *Laender*): number of bathers, management (e.g. presence of swimming instructors), 'an international or European interest' defined as at least 5,000 persons per day in season or so many per square meter. The local authorities drew up lists from which the *Laender* or regional authorities chose the areas. All the coastline has been identified. A total of 95 areas were identified in 1980, the majority of which were lakes, but the number has varied a little from year to year. For example, in Sarre the number dropped from five to two after 1980 and in Norther Rhine Westphalia four were suppressed and two new ones created. There are two transfrontier areas: Lake Constance, and one in the Baltic where measures are taken in common with Denmark by a joint committee. In nearly all cases the official reports to the Commission show that the standards were met and in no case have special measures had to be taken to meet the standards, since the necessary measures had been taken before the Directive. Problems in respect of the microbiological parameters exist with Lake Constance. In Lower Saxony some may need a waiver under Article 8. No derogations have been made. Many waters not identified under the Directive are used for bathing, some of which meet the standards while some do not. In some cases bathing is prohibited but the prohibition is not always respected.

The general effect of the Directive is to have increased knowledge about the quality of water rather than to have raised its quality. Much more sampling has been carried out and the Directive is criticised for requiring excessive reporting. The point is also made that it is easy to distort the true picture water quality by only giving global information in reports.

# United Kingdom

## FORMAL COMPLIANCE

The Control of Pollution Act 1974 (Part II) requires all discharges to inland, coastal and estuarial waters to be authorised. However, Part II is only now coming fully into effect and the previously existing legislation — the Rivers (Prevention of Pollution) Acts 1951 and 1961 — did not cover coastal waters nor all estuaries. However the British Government satisfied itself that it had the necessary powers for complying with the Directive since sewage that is the major threat to bathing waters is managed by the water authorities. However the Department of the Environment did not issues a circular appointing the water authorities as the competent authorities and setting guidelines for the identification of bathing waters until July 1979 — 18 months after the date for compliance. A list of identified bathing waters was sent to the Commission in December 1979 but because it did not indicate positively that no waters had been identified in Scotland and Northern Ireland the Commission issued a 'Reasoned Opinion' to the effect that the Directive was not fully complied with. The Commission has since stated in reply to a question in the European Parliament that it is not satisfied that the Directive is complied with. It seems that the Commission's dissatisfaction relates to the small number of bathing waters identified rather than with the legal provisions.

## EFFECT ON PRACTICE

Five water authorities have identified a total of only 27 bathing waters, none of which are fresh waters. The small number has been much criticised particularly as it excludes well known resorts such as Brighton and Blackpool. In fact it has been suggested that the criteria set down by the Department of the Environment (at least 500 people in the water at any one time, or 1,500 per mile) was chosen to exclude Blackpool which would have required an expenditure of between £10 to £50 million to meet the standards. Certainly the criteria were influenced by the Government's decision to hold down public expenditure. Several of the 27 identified waters did not meet the standards and in 1981 the Government informed the Commission that it was granting derogations in respect of four of the waters accompanied by management plans costing tens of millions of pounds. However some of these investments were programmed anyway, although the Directive has provided pressure for the schemes to be completed. The Department of the Environment is confident that the other waters will meet the standards by 1985. The greatest effect of the Directive is that it is being used as a standard when planning future investments in sewage treatments plants or long sea outfalls even to waters which are not identified as bathing waters.

# General assessment

All four countries were late in formally complying with the Directive. The Netherlands needed new legislation and was taken before the European Court; France was late with a new *arrêté*; the UK was late with a circular, and some of the *Laender* were late with their circulars. The Netherlands remains the most behind in still not having produced a list of bathing waters. It is now over five years since the list should have been sent to the Commission.

The manner in which waters have been identified varies in each of the four countries and problems arise in each:

- *Netherlands:* it seems that very flexible criteria are being applied and that fresh waters are not being identified if the quality is low. This should result in the prohibition of bathing but if it does not, it means that the Directive is being applied inconsistently. If the whole coastline is identified then there will be problems in meeting the standards in some places.
- *France:* an impressively large number of waters have been identified but a significant proportion (approximately 25 per cent) do not meet the standards and no derogations were submitted to the Commission. It is certain that some standards will not be met by the due date.
- *Germany:* all the coastline has been identified but the majority of identified waters are lakes. The list of fresh bathing waters has varied from year to year.
- *UK:* very few bathing waters have been identified quite deliberately in order that strains are not placed on public expenditure. The result is that many waters used for bathing are not included, but at least some consistency has been applied to the task of identifying bathing waters.

The Directive can be seen to place Member States in a dilemma: either they identify many waters (the course adopted by France) in which case many may not meet the standards by the due date and great expenditure will have to be incurred, or they can identify only a few waters (the course adopted by UK) to hold down public expenditure in which case the Directive does not achieve its intended purpose. Yet another course (which appears to be that being followed in the Netherlands for fresh waters) is for the Government to allow the water authorities to apply their own criteria for what is 'a large number of bathers' with the result that the Directive may be applied inconsistently as different authorities try to avoid the problems resulting both from the French course and from the British course. The position in the FRG is closer to that being followed in the Netherlands than in the other two countries.

The Directive has probably increased knowledge of bathing water quality in all countries and at least in France and UK has applied some pressure for expenditure on improvement works. It has also provided a reference standard which can be used even where waters are not identified under the Directive. Because the Directive gives inadequate guidance on the choice of sampling points and on how samples are to be handled before being analysed, it is not possible to make useful comparisons between the quality of water in different countries. It is significant that the Commission's consolidation of national reports on the quality of bathing water does not attempt to do so.

# 3.8 DANGEROUS SUBSTANCES IN WATER — DIRECTIVE 76/464
## No date set for formal implementation

## Brief summary of Directive

This Directive is the most significant concerned with water pollution. It covers inland, coastal and territorial waters. Discharges containing any dangerous substances listed in *List I* and *List II* are to be subject to prior authorisation but these authorisations are arrived at in different ways. *List I* substances are in general more dangerous than *List II* substances.

For controlling *List II* substances Member States are to establish pollution reduction programmes with deadlines. All discharges liable to contain a List II substance require prior authorisation with emission standards being laid down for the substances. These emission standards are to be based on quality objectives which must respect any existing Directives.

For controlling *List I* substances Member States may chose between two regimes. The preferred regime entails limit values for the substances (which emission standards are not to exceed) fixed in 'daughter' Directives. The alternative regime is like the regime for List II except that the quality objectives are also to be laid down in 'daughter' Directives. (All Member States except the UK have declared that they will adopt the preferred regime.) There is also a provision for inventories to be drawn up and for comparisons of the programmes for List II substances.

## Netherlands

### FORMAL COMPLIANCE

Since the 1969 Surface Waters Pollution Act all discharges have required prior authorisation. The 1981 amendments to the Act enabled limit values for discharges of designated substances to be set by statutory instrument but they also introduced a simpler procedure: a ministerial decree can be used to set limit values specifically for implementing the provisions of international agreements (this power has been used for the first 'daughter' Directive — see 3.10). These two powers are not limited to List I and List II substances but it is present policy only to use them for List I substances.

No specific provisions exist for the setting of national limit values for List II substances. Indeed List II substances are not recognised in Dutch policy as a separate category, though the setting of emission standards for List II substances is within the discretion of the water authority. The water quality plans to be drawn up under the Surface Waters Pollution Act by the provinces and by the Minister constitute the pollution reduction programmes. The explanatory memorandum to the Water Quality Plans Decree of 1982 made it clear that the minimum quality objectives to be laid down should follow the previously existing 'basic quality' criterion where no specific function is designated. Water quality objectives for the six priority List II substances identified by the Commission (chromium, lead, zinc, copper, nickel and arsenic) are included in the 'basic quality' parameters of the Indicative Multi-Year Programmes of 1975–1979 and of 1980–1984. However, there is no obligation on the water authorities to set quality objectives for all List II substances and the Minister does not have reserve powers to order the water authorities to perform this function. There is therefore a failure of formal compliance.

## EFFECT ON PRACTICE

In practice emission standards have been widely applied to discharges of List I and List II substances by way of licence conditions for many years but by no means in every case. For example it is still not yet general practice to include emission standards for List II substances in the licence for sewage works. Many sewage works where the operator is also the management authority for the receiving water have yet to be granted a licence under the Surface Waters Pollution Act. The Directive must be applying some pressure to clear this backlog, but its greatest effect is by increasing the role of central Government in laying down national emission standards for List I substances.

The requirements of the Directive for List II substances are consistent with the policy that previously existed in the Netherlands that discharges should be controlled with the aim of attaining a basic water quality. Following a meeting of experts from Member States organised by the Commission in January 1981 at which it was decided to compare national programmes for six priority List II substances a report was drawn up which summarised existing law, policy and practice. This report says that the control of List II substances is based on the best practicable means with more effective techniques being required as necessary to meet the relevant water quality objectives. This approach is confirmed in the draft water quality plans published in 1984.

# France

## FORMAL COMPLIANCE

A decree of February 1973 made under the Water Law of 1964 requires all discharges to be authorised and includes other provisions. This decree was supplemented by one of June 1976 dealing with sewage discharges and by *arrêtés* of May 1975 specifying criteria for authorisation. All these were reinforced by a circular of January 1977. The setting of authorisations for *Installations Classées* is done by reference to what is technically possible in economically acceptable conditions as determined by the *Inspecteur des Installations Classées* but must also respect quality objectives. Other discharges must respect quality objectives. The pollution reduction programmes are shown in water quality maps in accordance with a circular of March 1978.

## EFFECT ON PRACTICE

Discharges from *Installations Classées* of List I and List II substances are subjected to emission standards fixed on a case by case basis following standards proposed by the *Inspecteur des Installations Classées*. Industrialists lack confidence that their competitors in other countries are subject to equally stringent emission standards so that the idea of limit values fixed at Community level is welcomed. The Directive has had the effect of ensuring that List I and List II substances are now included in the standards proposed by the *Inspecteur* and the existence of the Directive has strengthened the hand of the authorities when negotiating with industrialists.

It is possible that not all discharges of List I and List II substances are individually authorised for existing plants and sewage works.

The requirements of the Directive for List II substances are consistent with the French law and practice that emission standards must be set by reference to quality objectives. Following the meeting convened by the Commission in 1981 on List II substances, France selected chromium, lead and arsenic for particular attention and quality objectives are being elaborated accordingly.

# Germany

## FORMAL COMPLIANCE

The development of this Directive coincided with preparatory work leading to the amendment in 1976 of the law on water and played a key role in German water pollution policy. Previously policy was based on the quality of the receiving water and the authorities had considerable discretion in authorising discharges. In 1975, the *Laender* working group on water (LAWA) agreed a short paper known as the 'Mainzer Papier' which put the emphasis on minimum mandatory emission requirements based on the state of technology with the possibility of more stringent standards to meet a given water quality standard. However, it recommended against uniform environmental quality standards for all waters with the same use. The 'Mainzer Papier' thus forsaw a double approach both by 'emission' and by 'immission' (e.g. a reference to the quality of the receiving water). It influenced the amendment of the water law in 1976 since the emission approach did not exist in earlier versions of the law. Emission standards, which according to the 'Mainzer Papier' were to be fixed on the basis of the state of technology, are according to the 1976 law to be fixed on the basis of 'generally recognised technical rules'. The phrase in the Directive 'best technical means available' goes further in requiring that the best means be sought. The German interpretation of 'generally recognised technical rules' includes economic feasibility as does the interpretation put on the Directive by a Council minute. Thus in Germany the Directive is seen to correspond closely to German laws and is generally thought to be the most important water Directive. The 1976 law provides powers for the authorisation of all discharges without a distinction between List I and List II substances, and the Federal Government can issue regulations setting minimum requirements for discharges for specific branches of industry. Some gaps in the Federal law (which is only a framework law) are closed by the laws of the *Laender*, e.g. the Federal law says nothing about discharges to sewers. The 1976 law also requires management plans for water under which 'immission standards' can be set. (This German terminology can be understood as 'quality objectives'.)

## EFFECT ON PRACTICE

Minimum requirements for emission of pollutants, covering the List I and List II substances, are fixed in Federal regulations dealing with specific branches of industry. About 50 are expected of which 45 have so far been issued. Where no minimum requirements have been set, the competent authorities set their own emission standards. Not all old plants have authorisations specifically referring to discharges of List I substances, but new ones do.

Following the meeting convened by the Commission in 1981 the Federal Government has prepared pollution reduction programmes for the six priority List II substances selected by the Commission by indicating the quality standards of Directive 75/440 on surface water (see 3.2). These are applied also to coastal and estuarial waters.

# United Kingdom

## FORMAL COMPLIANCE

This Directive has played a key role in the development of British water pollution control policy but has not yet affected British legislation. The Control of Pollution Act 1974 (Part II) gives powers to the water authorities to authorise discharges to inland, coastal and estuarial waters and groundwater, but it is only now coming fully into effect and under the previous legislation — Rivers (Prevention of Pollution) Act 1951 to 1961 — not all discharges to coastal waters and estuaries were controlled. For the last few years there has therefore been a failure of compliance. The water authorities have a duty under the Water Act 1973 to prepare plans for improving water and can do this by seting river quality objectives. No distinction is made in the legislation between List I, List II and any other substances.

It was the UK that objected to limit values for List I substances which resulted in the Directive having two alternative regimes for List I. Since the UK has decided not to adopt the approach involving limit values for List I substances the fact that there are no formal powers to set these limit values nationally has not presented any problem.

## EFFECT ON PRACTICE

The major effect of the Directive has been as much a refinement of thought as a change in practice. The British have been stimulated into developing their previously imprecise ideas on the use of environmental quality objectives since the Directive makes these mandatory for the first time. River Quality Objectives were laid down in 1978 for all stretches of river according to use but these did not include numerical standards for all List I and List II substances. Accordingly research was undertaken by the Water Research Centre (WRC) relating to six priority List II substances: chromium, lead, zinc, copper, nickel and arsenic. In October 1984 the Department of the Environment issued a circular saying that the WRC recommendations should be adopted as national quality standards, and the UK was therefore the first Member States systematically to introduce environmental quality standards for List II substances for a number of water uses. The difference between these new standards for List II substances as compared with the quality objectives that have previously existed in UK, in other EC Directives, and in other countries such as the Netherlands and France, is that numerical standards have been laid down for a number of different uses for each substance. (In the previous EC Directives numerical standards are laid down for a number of substances for one water use. The water quality maps in existence in France do not yet cover all waters. In the Netherlands a single 'basic quality' with quality standards for the six substances has existed since 1975 but separate water functions or types of water were not recognised.)

Although powers exist for authorising discharges of List I and List II substances not all discharge consents set emission standards for each List I and List II substance individually. For example, all sewage works discharge some List II substances (e.g. zinc) but emission standards for zinc will not necessarily have been laid down.

# General assessment

This Directive has had a significant effect on water pollution policy in Netherlands, Germany and Britain and has had some effect in practice in France. However, the effects have been quite different in the different countries. In the Netherlands new legislation had to be introduced and central government is now involved for the first time in laying down national emission standards for List I substances. In Germany the development of the Directive coincided with the development of new federal legislation and reinforced the move away from locally determined emission standards set by reference to the receiving waters towards emission standards centrally fixed in accordance with available technology. In Britain the Directive had the opposite effect and hardened opinion against centrally fixed emission standards, both on the grounds that this approach was theoretically incorrect but also because of economic self-interest. This self-interest stems from the fact that in Britain, unlike the other three countries, it is much rarer for polluting industries to discharge into long slow running rivers which might also be used for drinking water supply. Instead they discharge to estuaries or the sea. Having opposed centrally fixed emission standards, the British have been forced to show their commitment to the alternative regime and this may explain why they are further advanced with setting quality standards for List II substances. However, in the Netherlands quality standards have existed for List II substances at least since 1975 and the principle has also existed in France.

In France the Directive has had the effect of ensuring that List I and List II substances are now included in the standards proposed by the *Inspecteur des Installations Classées* and has also reinforced the hand of the authorities when negotiating with industralists.

In all four countries not all old discharges containing significant quantities of List I and List II substances have emission standards referring individually to the substances. This is particularly true of sewage works. In fact the approach in all four countries had previously been to set emission standards more by reference to the type of plant rather than by individual substances so that the Directive has concentrated attention on the List I and List II substances.

# 3.9 GROUNDWATER — DIRECTIVE 80/68
## Date for formal implementation — 19.12.81

## Brief summary of Directive

This Directive has subtle and complicated provisions which are sometimes imprecise although its principles are clear. All direct discharges (i.e. without percolation through the ground) of *List I* substances are to be prohibited (except in trace quantities), though if after investigation the groundwater is found unsuitable for other uses such discharges may be authorised. All direct discharges of *List II* substances are to be subject to investigation before being authorised.

Any disposal to land of *List I* or *List II* substances which might lead to indirect discharges is to be subject to investigation before being authorised (authorisations are also required for waste disposal which often gives rise to indirect discharges — see Sections 5.1 and 5.2). Any other activity likely to lead to indirect discharges of *List I* substances is to be controlled. Artificial discharges for the purpose of groundwater management are to be specially authorised on a case by case basis. All authorisations may only be issued if the groundwater quality is undergoing the requisite surveillance.

## Netherlands

### FORMAL COMPLIANCE

Control over groundwater is presently fragmented in a wide variety of unrelated Acts (e.g. Nuisance Act, Chemical Wastes Act) so that only incidental elements of the Directive are formally implemented. As a result, the Commission initiated infringement proceedings against the Netherlands in 1983. The Ground Protection Bill currently before parliament should allow implementation although various statutory instruments will be necessary under the Act to complete the implementation process.

## EFFECT ON PRACTICE

Not a single change to Dutch policy and practice can be attributed to its provisions except for the pressure to introduce the Ground Protection Bill — although some form of Bill will have been introduced anyway without the Directive. The explanatory memorandum accompanying the Bill makes it clear that the primary influences were the groundwater Directive, the dangerous substances Directive 76/464 (see Section 3.7), the Rhine Convention on chemical pollution and the Council of Europe's 1973 Soil Charter.

# France

## FORMAL COMPLIANCE

Two Decrees of 1973 (73/218 and 73/219) made under the Water Law of 1964 require all direct discharges to groundwater to be subject to authorisation. Discharges, both direct and indirect, from *Installations Classées* (including some waste disposal sites) also have to be authorised under the relevant law of 1976 and direct discharges are prohibited. A circular of November 1983 requires all unauthorised waste disposal sites to be closed or to be subjected to authorisation.

In addition, a procedure for protecting aquifers for drinking water exists under the *Code de la Santé Publique*. Boundaries are defined by the authorities based on expert geological advice and, within these boundaries, restrictions are placed on activities required to protect groundwater. In connection with this a circular is to be issued in 1985 concerning the boundaries and restating all the requirements of the Directive. This has been delayed because of the new law on decentralisation which now gives to the *maires* certain powers over planning matters including the setting of the boundaries. The Commission has not been satisfied that formal compliance is complete and started infringement proceedings in 1983 although the French authorities believe that all the formal requirements of the Directive are to be found in existing French measures.

## EFFECT ON PRACTICE

Procedures to protect groundwater were already in existence before the Directive, but not all indirect discharges were controlled since not all waste disposal sites were authorised. Programmes of control and surveillance of groundwater quality introduced in nearly all *départements* show that the condition of groundwater is not always good. Although the Directive is less well known than the other water Directives it has probably helped to reinforce existing French preoccupations.

# Germany

### FORMAL COMPLIANCE

In December 1981 the Federal Government informed the Commission that five *Laender* had satisfied the requirements for formal compliance with the Directive and in July 1984 a full communication was sent to the Commission listing the measures adopted by all 11 *Laender*. Most of the measures were circulars based on a common model. All these measures depend from the federal law of 1976. Whereas these measures are adequate for controlling direct discharges of List I and List II substances, doubts exist about formal compliance with the requirements of the Directive over indirect discharges.

### EFFECT ON PRACTICE

An extensive range of measures have been taken in Germany to study and prevent the pollution of groundwater and it is doubtful that the Directive has had any clear effect in practice. However there is considerable doubt that indirect discharges are adequately controlled and the Directive has helped to focus attention on this problem. This is despite the fact that Germany played a major role in the development of the Directive.

# United Kingdom

### FORMAL COMPLIANCE

The Control of Pollution Act 1974 (Part I) gives the waste disposal authorities powers to grant or withhold authorisations for the direct or indirect discharges from waste disposal sites, and Part II of the same Act gives powers to water authorities to authorise or withhold authorisations for the direct or indirect discharge of effluent to groundwater. The relevant sections of Part II are only coming into effect in 1985 although previously existing legislation gives some of the necessary powers. There has therefore been a delay in formally complying fully with the Directive. Some powers under the Town and Country Planning Act 1971 enable the disposal of waters from mines and quarries to be controlled.

In 1982 the Department of the Environment issued a circular explaining the Directive to the waste disposal authorities, water authorities and mineral planning authorities and appointing them the competent authorities for the discharges under their control.

**EFFECT ON PRACTICE**

The major advance in control of groundwater in Britain followed the creation of the water authorities in 1974 and the bringing into force in 1976 of Part I of the Control of Pollution Act which requires waste disposal authorities to consult the water authorities before authorising a waste disposal site. Several water authorities have drawn up aquifer protection policies indicating zones where waste disposal is unacceptable. None of this can be attributed to the Directive. What the Directive has done is to concentrate attention specifically on the possibility of the List I and List II substances reaching groundwater either from waste disposal sites or elsewhere, although the DOE believes that the number of waste disposal sites affected is likely to be very small. Some minor changes to administrative practice can be attributed to the Directive.

# General assessment

In the Netherlands 90 per cent of drinking water is taken from groundwater. In France the figure is 60 per cent, in Germany 80 per cent and in UK 25–30 per cent. In all four countries measures have been taken to protect groundwater before the Directive existed, and since the Directive is imprecise in many respects and contains exceptions to meet the practices existing in certain countries it is hard to point to specific changes in practice resulting from the Directive. Its greatest effect has probably been to emphasise and reinforce existing policies. In the Netherlands it has had the further effect of influencing the Ground Protection Bill.

# 3.10 MERCURY FROM THE CHLOR-ALKALI INDUSTRY — DIRECTIVE 82/176
## Date for formal implementation — 1.7.83

# Brief summary of Directive

As provided for in the parent Directive 76/464 (see Section 3.8) the authorisations for discharges of mercury from chlor-alkali electrolysis plants (which produce chlorine) are to conform either to specific *limit alues* or *quality objectives*.

he limit values are expressed both in terms of concentration ($\mu$g/l of lischarges) and in terms of quantity (grammes per tonne of chlorine apacity). Different limit values in terms of quantity apply to the recycled rine and lost brine processes. The limit values in terms of quantity must e observed and in terms of concentration should be respected.

our quality objectives are laid down (fish flesh, inland surface water, stuary waters, sea and coastal waters) but it is for the competent authority ɔ select the one that is appropriate.

special provision relates to new plant (i.e. built or modified after March 982) intended to ensure that the best technical means are used.

# Netherlands

## ORMAL COMPLIANCE

he 1981 amendment to the Surface Waters Pollution Act made it possible ɔ lay down national limit values for discharges of designated substances see Section 3.8). The first use of this provision was to implement this irective. A Command was issued in June 1983 making the limit values rovisions of the Directive effective from 1st July 1983. Only values for the ecycled brine process are laid down since this is the only kind operating 1 the Netherlands.

## FFECT ON PRACTICE

here are two plants in the Netherlands, both discharging to inland waters. oth plants previously had licences for emissions which are equivalent to 1e limit value of the Directive so that the Directive has had no practical ffect. The quality objectives approach is not being followed although the 1dicative Multi-Year Programme for 1980–1984 includes a quality bjective as part of the 'basic quality' criterion for all surface water of 0.5 g/l, i.e. half that set in the Directive and the quality objectives of the irective are met.

# France

## ORMAL COMPLIANCE

he law of 1976 on *Installations Classées* and its *Décret d'Application* of

September 1977 enable discharges from the chlor-alkali industry to be controlled in accordance with the Directive. Before the Directive a technical instruction of 1974 specified standards for discharge of mercury but this was modified by a circular of December 1983 which specified the limit values of the Directive. Formal compliance was therefore six months late.

## EFFECT ON PRACTICE

There are seven plants in France listed in an annex to the circular of December 1983. All these discharges are into rivers except one into the Mediterranean near Marseilles. This annex shows that these plants did not all meet the limit values in 1983. The circular also stated that the limit values of the Directive were not always respected and said that the necessary action should be undertaken as a matter of urgency. Since 1974 steps had already been taken to reduce emissions and the Directive is having the effect of applying new pressure for further reductions.

# Germany

## FORMAL COMPLIANCE

The federal law of 1976 provides the powers for implementing the Directive and in June 1984 standards were laid down in a federal regulation specifically dealing with the chlor-alkali electrolysis industry. The regulation is based on the limit values of the Directive but adapts them to German practice. At the federal level formal compliance was a year late, but the six *Laender* where plants are located had already issued circulars prescribing the limit values of the Directive although some of these circulars were also late. These circulars are almost identical and go further than the federal regulation by fixing the standard for concentrations which is not obligatory under the Directive. The quality objective approach is not mentioned in the Federal regulation but the circulars of the *Laender* specifically say that the quality objective approach is not to be followed.

## EFFECT ON PRACTICE

There are 13 plants in Germany. The emission standards set by the authorities are in all cases as stringent as those in the Directive, and sometimes more stringent. They are respected in practice, one case when the standards were exceeded being the result of changes in the rhythm of production. The frequency of sampling is slightly different from that in the Directive since samples have to be taken every two hours in theory and the limit value is said to be respected if the average of the five previous samples taken by the control authority does not exceed it. To compensate the German limit values are more severe than in the Directive.

# United Kingdom

## FORMAL COMPLIANCE

The Control of Pollution Act 1974 requires all discharges to water to be authorised. There is no specific power to set emission standards nationally but the Government can advise the water authorities on certain standards and can, if necessary, compel them to adopt them. The water authorities can set quality objectives administratively and can be compelled to adopt them. In August 1982 the Department of the Environment issued a circular appointing the water authorities as competent authorities and instructing them to apply the Directive. The circular said that it was the Government's intention that the Directive would be applied using the quality objective approach. The Government submitted a case to the Commission for the use of this approach under Article 6.3 of Directive 76/464 in July 1983.

## EFFECT ON PRACTICE

There are five plants in the UK, one discharging to an estuary, three into canals and one into a river. Four of these plants have authorisations set by reference to the quality objectives of the Directive although one of these plants will not meet the quality objective until 1986. The fifth — the Sandbach Works in Cheshire (now owned by Hays Chemicals, although built by BP Chemicals) — discharges into the Trent-Mersey canal and the environmental quality objectives of the Directive are not yet met so that the authorisation has been set in accordance with the limit values of the Directive. This is an obvious inconsistency in the British position but it remains the intention of the authorities that the quality of the canal (which is also affected by mercury in the sediment deposited in the past) will eventually be raised to that set in the Directive. This plant was redesigned after Directive 76/464 was agreed but before the standards for mercury were set in Directive 82/176. The knowledge that the Directive was coming ensured that the best available technology was used and provides an example of the prospect of a Directive creating pressure on an industralist.

It is for the competent authority to select the appropriate quality objective from among those listed in the Directive and also to determine the area affected in each case. In some cases, the areas selected include: the canal into which the plant discharges, an estuary into which the canal discharges, and a stretch of coastal waters (e.g. Liverpool Bay). The objectives selected and areas affected have been communicated to the Commission. The process of selecting the areas and monitoring to ensure that the objectives are met has involved extra work and has increased knowledge about mercury concentrations and its sources including sources other than from the chlor-alkali industry. By 1986 it is expected that only one plant (six per cent of UK chlorine production capacity) will not meet the limit values.

# General assessment

This is the first 'daughter' Directive under Directive 76/464 and is therefore the first occasion for putting to the test the compromise arrangement whereby Member States may choose not to follow the approach based on limit values for emission standards, and instead to adopt the quality objective approach. As expected, only the United Kingdom has chosen in principle to follow the quality objective approach.

In the Netherlands the two plants concerned both meet the limit values, as do the plants in Germany. In France some of the seven plants do not yet meet the limit values so that the Directive is having the effect of applying pressure for improvement. There are five plants in the UK, four of which are authorised in accordance with the quality objective approach, while the fifth plant is authorised to meet the limit values since the quality objective is not yet met. The theoretical approach adopted by the UK is not therefore being followed in a consistent manner and can only increase awareness of the need to integrate the use of the two approaches. It is significant in this connection that although the quality objective approach may sometimes result in more relaxed emission standards than the limit value approach, it can sometimes result in more stringent emission standards than the limit value.

The above descriptions of the practices in the Member States show that the limit value approach is easier to apply administratively since it does not involve selecting the appropriate quality objective, determining the areas affected and monitoring all the affected waters. It also has the advantage that industralists know that all plants are treated equally. The quality objective approach however has the advantage that there is more extensive monitoring of the environment and that sources of mercury other than from the chlor-alkali industry are taken into account.

# 3.11 TITANIUM DIOXIDE — DIRECTIVES 78/176 AND 82/883
## Date for formal implementation — 22.2.79

## Brief summary of Directive

Titanium dioxide ($TiO_2$) is a white pigment the manufacture of which usually results in much larger quantities of waste than product. All discharge, dumping, storage and injection of waste must be subjected to prior authorisation. The information to be supplied with a request for authorisation is specified and Directive 82/883 specifies monitoring procedures. Pollution reduction programmes (leading to eventual elimination of pollution) were to have been sent to the Commission by 1980 including targets to be achieved by July 1987. The Commission could agree to a request from a Member States made before August 1979 (Article 10) that no pollution reduction programme was necessary for a particular plant.

The Commission is to submit proposals for harmonising the national pollution reduction programmes (this has been done — COM(83)289 — but the Directive has not yet been agreed). Reports are to be sent to the Commission.

## Netherlands

### FORMAL COMPLIANCE

Discharges to inland, coastal and territorial waters are controlled under the Surface Waters Pollution Act, and although there are no specific provisions covering discharges from $TiO_2$ plants the authority is empowered to demand all relevant information when granting a licence. Pollution reduction programmes can be included in the general water quality plan. All dumping in territorial waters or the loading of a ship with waste for dumping inside or outside territorial waters requires an exemption from the general prohibition contained in the Seawater Pollution Act 1975. Exemption is granted by the Minister of Transport and Public Works and may be subject to conditions. The Act came into effect in 1977 but the exemption for the dumping of $TiO_2$ waste by two German companies (see below) was only given in 1980.

### EFFECT ON PRACTICE

There are two sources of $TiO_2$ waste needing disposal in the Netherlands:

there is the Tiofine works using the sulphate process in Rotterdam (the only Dutch $TiO_2$ producer) and waste from two German companies (see below) which is dumped at sea off the Hook of Holland.

The Tiofine plant is situated on the Nieuwe Maas downstream of Rotterdam and about 16 kilometres from the North Sea. It has been a major source of concern for many years. The plant was granted a new licence in 1979 by the Minister of Transport who, as the licensing authority of the only $TiO_2$ plant in the Netherlands, was directly able to implement the provisions of the Directive. Conditions attached to the licence required the discharges of acidic waste to be halted within two years with the possibility of two one year extensions. The Minister intended the licence to constitute the pollution reduction programme and it was sent to the Commission in January 1981; however, the timetable was disrupted by appeals against the licence by Tiofine itself claiming the conditions were too onerous and by an environmental group claiming that the period was excessive and that the monitoring requirements did not comply with Directive 78/176. In its judgement of August 1984 the Crown approved a new licence, to be valid until 1st January 1988, and stipulated that an improvement plan for ending the discharges of acidic wastes was to be submitted within one year and be put into effect within three years.

The acidic waste from the two German plants is shipped down the Rhine by barges for loading onto a dump ship and was first authorised in 1980. The current exemption was granted in 1984 for two years but the Minister declared that further dumping would be permitted until 1989 if the two companies proceeded with their pollution reduction programmes (which are a German responsibility — see below). The issues have been complicated by litigation.

Despite the delays occasioned by the litigation it is nevertheless clear that the Directive has put pressure on the Dutch government to take a tougher line with the titanium dioxide industry, and any future reductions in acidic wastes can be largely attributed to the Directive.

# France

## FORMAL COMPLIANCE

The legislation on *Installations Classées* provides all the powers for formal compliance with the Directive, and $TiO_2$ production is specifically listed as coming under the legislation.

## EFFECT ON PRACTICE

There are three $TiO_2$ plants in France:

- Thann-Mulhouse (sulphate process) discharging at Le Havre into the estuary of the Seine
- Tioxide (sulphate process) discharging to sea at Calais through a 1,300m pipe
- Thann-Mulhouse (sulphate process) at an inland site at Thann which discharges into the River Thur.

Improvement programmes were submitted to the Commission in November 1980 and these programmes were the direct result of the Directive.

# Germany

## FORMAL COMPLIANCE

The competence for waste disposal from $TiO_2$ production lies with the Federal Ministry of the Interior. No extra legislation or administrative measures have been required for formally complying with the Directive. Only two *Laender* are involved with the execution of certain functions (Lower Saxony and North-Rhine Westphalia). The Federal Hydrographical Institute is responsible for authorising dumping at sea.

## EFFECT ON PRACTICE

There are four $TiO_2$ plants in Germany:

- Bayer (sulphate and chloride process) at Krefeld Uerdingen whose wastes are recycled although some weak acids are discharged
- Kronos Titan (sulphate and chloride process) at Leverkusen whose wastes are transported by barge and dumped at sea
- Kronos Titan (sulphate process) at Nordenham whose wastes, including copperas, are transported by barge and dumped at sea
- Sachtleben Chemie (sulphate process) whose wastes are transported by barge and dumped at sea.

In 1978 the Federal Government submitted under Article 10 that no pollution reduction programmes were necessary but the Commission only accepted the case for Bayer. In July 1980 pollution reduction programmes were submitted to the Commission proposing complete cessation of sea dumping of copperas in 1987. In October 1983 this programme was supplemented by another according to which the dumping at sea of weak acid would cease in 1989 and the dumping at sea of copperas would cease in 1984.

The plans to stop dumping are the result of technological development but represent a complete change of position on the part of the Federal Government which, during negotiations on the Directive, argued that dumping at sea was perfectly acceptable. The Directive has played a major part in this change of attitude.

# United Kingdom

## FORMAL COMPLIANCE

The Control of Pollution Act 1974 contains all the powers for authorisation to water in accordance with the Directive, but it is only now coming fully into effect. As one British $TiO_2$ plant was not covered by the previously existing legislation it was not formally authorised until recently so there was a formal failure of compliance. Some neutralised waste is dumped on land and is authorised under the part of the Control of Pollution Act concerned with waste. There is no dumping at sea.

## EFFECT ON PRACTICE

There are three $TiO_2$ plants in Britain:

- Tioxide at Hartlepool (chloride process) discharging into the estuary of the Tees
- Tioxide at Grimsby (sulphate process) discharging into the estuary of the Humber
- SCM Corporation (who bought the plant from Laporte Industries Ltd) at Stallingborough (sulphate and chloride processes) discharging into the estuary of the Humber.

In 1978 the Government submitted under Article 10 that no pollution reduction programmes were necessary for the discharges to the Humber but the Commission rejected the submission. The companies did not accept the Commission's opinion and brought a case before the European Court to annul the Commission's opinion although they have not proceeded with it.

A pollution reduction programme was submitted to the Commission for the Tees in July 1980 and for the Humber in January 1981. These programmes did not show how quality objectives for the two estuaries were to be achieved by the measures taken despite the fact that the British Government had, during negotiations on the Directive, resisted the proposal for a fixed percentage reduction in discharges on the grounds that what mattered was whether the quality of the receiving environment was satisfactory. Some elements of the pollution reduction programmes are not being proceeded with until it becomes clear what form the proposed 'harmonising' Directive — COM(83)289 — finally takes. The Directive has had the effect of focussing attention on discharges which previously have not been considered particularly harmful given the location of the plants. It will also have been a factor in the decision to expand production by the much cleaner chloride process rather than the sulphate process.

# General assessment

The Directive arose because of controversy surrounding the dumping into the Mediterranean of waste from a plant in Italy in the early 1970s, and the dumping of German waste in the North Sea from ships from the Netherlands continues to cause controversy. The requirement of the Directive for pollution reduction programmes has provided a focus for discussion of sea dumping between the Netherlands and Germany and by opponents of dumping. This can be seen from the fact that exemption from the ban on dumping granted by the Netherlands until 1989 is based on the German pollution reduction programmes. The German decision to stop dumping altogether is a striking reversal of the position it adopted during negotiations on the Directive.

In France, and Britain, the Directive has encouraged improvements to methods of discharging but any further major changes are likely to be delayed until the 'harmonising' Directive is agreed. In the Netherlands improvements to the Tiofine plant — to which the Directive gave an impulse — were held up by a Court case. In Britain the increasing use of the chloride process is at least partly attributable to the Directive.

In Britain also the Directive has illustrated the difficulties surrounding the questions raised in the continuing dispute about rival theories of pollution control. Should industrialists be allowed to benefit from the comparative advantage brought them by geography, or should the same technical solutions be adopted by all manufacturers irrespective of the ability of the environment to accept more waste in some places than in others? The practical difficulties of defining useful quality objectives for estuaries has been highlighted by this Directive. Discussions on the proposed 'harmonising' Directive — COM(83)289 — show that this debate is far from over.

# 4
# Waste — Legislation and Responsibilities in the Four Countries

There are four Directives concerned with waste. One sets a general framework for all waste and the other three make more detailed provisions for toxic waste generally, for PCB (polychlorinated biphenyls) and for waste oil. The framework Directive has three simple elements:

- there must be *competent authorities* responsible for waste disposal
- they must produce waste disposal *plans*, and
- they must *authorise* all waste disposal facilities.

These three elements exist in the legislation or administrative provisions of all four countries, but the arrangements differ quite significantly between countries and sometimes also differ within each country depending on whether household waste or toxic waste is being handled.

## NETHERLANDS

New arrangements for administering the disposal of domestic commercial and industrial wastes were introduced by the Wastes Act 1977. These wastes are divided into three categories — domestic refuse, derelict motor vehicles and 'other waste' (but not toxic waste which is covered by separate legislation — see below). Provisions for disposal vary respectively, though in each case the responsibility for co-ordinating disposal procedures lies with the provinces.

Each municipal council is obliged to collect domestic refuse from all premises in its district at least once a week and must issue local ordinances regulating the disposal of such waste. In co-ordinating waste disposal each provincial council must divide its region into 'co-operation areas' comprising groups of municipalities which are to work together as a unit. With regard to derelict motor vehicles, the Act prohibits the owner of such a vehicle from depositing it in a place visible to the public or to hand it over to anyone other than a licensed disposal establishment. 'Other waste' — principally industrial waste excluding toxic waste — may again only be disposed of through an authorised facility.

As the authority responsible for co-ordinating the disposal of waste, each province is required under the Wastes Act to draw up a disposal plan for domestic waste and a separate plan for industrial waste which is to be disposed of with domestic refuse. Additional plans for special categories of waste may also be required by statutory instrument. To date three such categories have been designated — construction and demolition waste, derelict motor vehicles and hospital waste.

The Wastes Act requires that the establishment or alteration of any waste facility is to be licensed by the provincial council. Any conditions attached to the licence are to be in the interests of environmental protection, and the licence itself must conform to the provisions of the provincial waste disposal plan.

The handling of wastes which pose particularly severe environmental problems is regulated by the Chemical Wastes Act 1976. Implemented in 1979, the Act introduced a blanket prohibition on the dumping of 'chemical waste' on land and prescribed a licensing system for its storage, working, processing or destruction. Chemical waste is defined as material which contains one or more of 83 designated substances and compounds in greater than specified concentrations plus all waste material from nine designated categories of industrial activity. The storage, working, processing or destruction of chemical waste requires a licence from the Minister of Housing, Planning and Environment. Any such wastes must be handed to a licensee for disposal, and both parties must supply the Minister with full details of the transaction. Conditions may be attached to the licence as appropriate. In certain circumstances the Minister may also grant an exemption to permit the waste to be dumped on land. There is no obligation on the Minister to produce a plan for the disposal of chemical waste.

In general controls over the disposal of waste oil follow those laid down for regulating toxic waste — there is a general prohibition on the dumping of waste oil on land and its collection, storage, working, processing and destruction is subject to a licence from the Minister.

# FRANCE

The Law on Wastes of July 1975 places the responsibility for collecting and disposing of household waste on the *communes* but foresaw that *communes* might group themselves together possibly in liaison with the *départements* to arrange for disposal of household waste. An interministerial circular of 18th May 1977 required the Prefects to create departmental working groups to put in hand departmental plans for disposal of household waste and to work out rules to be applied by the *communes* for the disposal of household waste. Thus, although the legal responsibility for disposal rests with the *communes*, in practice the disposal plans are effectively made at a departmental level.

Some landfill sites receiving certain industrial waste are *Installations Classées* (see below) and subject to the authorisation procedure laid down by the law of July 1976 on *Installations Classées*. All other landfill sites have to be authorised by the Prefect of the *département*. A large number of unauthorised sites existed at the time the 1975 law was published but a circular of 22nd November 1983 stated that all these sites either had to be authorised and made subject to the appropriate technical rules or had to be closed.

A circular of 22nd January 1980 concerned with disposal to land of industrial waste defined three categories of site:

- First class or impermeable sites
- Second class or semi-permeable sites
- Third class or permeable sites.

The authorisation for each site specifies a list of the waste that is forbidden on the site and the waste that is permitted.

The law of 1976 on *Installations Classées* and its *Décret d'application* of September 1977 specifies that transit stations, landfill sites, treatment facilities and incinerators receiving waste from *Installation Classées* are themselves *Installations Classees* and hence subject to a special authorisation procedure. The law of 1975 on wastes and its *Décret d'application* of August 1977 specify waste under five headings about which information must be furnished to the competent authority. These headings list 28 specified toxic substances, and waste from certain industries.

A circular of June 1980 recommended that *Commissaires de la République* of the Regions should create working groups to work out regional waste disposal plans for industrial waste. These are not obligatory and so far 13 regions out of 22 have established working groups. An *arrêté* of January 1985 requires toxic waste producers to follow their waste until disposed of and in particular requires an identification form to accompany toxic waste during transport.

At a national level it is the Ministry of the Environment that defines policy on waste management in liaison with other affected Ministries. For implementing the policy the Ministry relies on the *Directions Régionales de l'Industrie et de la Recherche* (DRIR) and on the *Agence Nationale pour la Récupération et l'Elimination des Déchets* (ANRED). ANRED is a national body, financed by the Ministry of the Environment and by the Ministry of Industry and Research, that exists to promote treatment facilities and recycling, to develop new techniques and to advise industries and local authorities. It can give subsidies for experimental purposes. The *Agences de Bassins* also have a role in financing water treatment and hence have some involvement with waste disposal.

# GERMANY

The Basic Law gives the Federation and the *Laender* concurrent legislative competence in the matter of waste disposal. This means that once the Federation has legislated the *Laender* are no longer free to do so. The Federation has used its powers with its law of elimination of waste — now in its 1977 version, although originally enacted in 1972. The *Laender* have transfered this law into their own legislation and have only been able to introduce original features where this is specifically provided for in the Federal law. The one feature where the *Laender* have freedom is in the appointment of the competent authorities. The 11 laws of the *Laender* therefore only differ from one another in this matter of competent authorities.

The laws regulate the elimination of waste, that is to say collection, transport, treatment, storage and disposal. Waste is defined and includes waste for which more severe regulations are necessary because they are hazardous. A Federal Government decree of 1977 lists these 'special' wastes. All waste disposal sites have to be authorised. Landfill sites are authorised under the law on wastes, incinerators are authorised under the law on classified installations. Producers of special waste are obliged to maintain a register of these wastes, and this obligation can also be imposed on producers of other non-domestic wastes. Collection, transfer, transport, the import, and (since 1984) the export of waste has to be authorised. Those involved in transport of waste from its production to its point of disposal must submit indentification forms to the competent authorities. Each *Land* must produce waste disposal plans.

A law on waste oils establishes a system of control and creates a fund from which payments can be made to authorised establishments for collecting and disposing of waste oils.

The Federal law requires each *Land* to designate the authorities responsible for the implementation of the law and also the public authorities responsible for waste disposal. The responsibility for authorisations is generally given to the *Kreis* and/or the regional authorities *(Regierungs Praesident)*. The responsibility for plans is a matter either for the Ministry of the *Land* or for the *Regierungs Praesident*. The responsibility for disposal is either a matter for the communes or for the *Kreis*, who may in turn delegate this task, may use the services of private bodies, or may form an association for the task. All the authorities may themselves establish waste disposal facilities which producers of toxic waste can be compelled to use.

# UNITED KINGDOM

The Local Government Act 1972 appointed the county councils in England and the district councils in Wales as the waste disposal authorities for all waste including toxic waste. In Scotland and Northern Ireland the district or island councils are the waste disposal authorities. Part I of the Control of Pollution Act 1974 requires all waste disposal authorities to draw up plans and all waste disposal sites to be licensed. The Act specifies the matter that must be included in the plans. The disposal licences can be made subject to whatever conditions the waste disposal authority considers appropriate, and the water authorities must be consulted before the licence is given. A public register with particulars of all disposal licences must be maintained.

Regulations made under the Control of Pollution Act defined 'special wastes' which are subject to extra controls because they are hazardous. A number of substances are listed and any wastes containing these in certain concentrations are 'special waste'. The definition is essentially by reference to its possible effect on human health. Producers of 'special waste' must notify the waste disposal authority in advance that waste is to be transported to a particular site and a consignment note must travel with the waste. Records must be kept of the despatch, transport and disposal of all 'special waste'. Chemical incinerators require a disposal licence but are also controlled by the Industrial Air Pollution Inspectorate.

At a national level the Secretary of State has power to direct acceptance and disposal of 'special waste' at a particular site or plant, but this power has not been used. A Chief Inspector of Hazardous Wastes was appointed in 1983 to advise the waste disposal authorities.

# GENERAL COMPARISON

In both Germany and UK the same competent authorities are responsible for both domestic waste and hazardous waste and the same legislation covers both. In both countries plans for hazardous waste can be included in the waste disposal plans.

In France the authorisations for both domestic and hazardous waste sites are granted at the level of the *département* but separate plans are prepared for domestic and hazardous waste: domestic waste disposal plans must be prepared by working groups of the *communes* formed in each *département*, whereas plans for industrial waste (including hazardous waste) are not obligatory but some have been prepared at regional level.

In the Netherlands hazardous waste is covered by quite separate legislation from domestic and other non-hazardous waste. Disposal of domestic and non-hazardous industrial waste is authorised by the provinces who also draw up plans. Authorisations for hazardous waste are however issued at the national level by the Minister himself. There is no obligation on the Minister to draw up a plan for hazardous waste.

# 5

# Waste Directives

## 5.1 WASTE — FRAMEWORK DIRECTIVE — DIRECTIVE 75/442
### Date for formal implementation — 18.7.77

### Brief summary of Directive

The Directive contains five fairly simple elements without going into much detail:

- there must be competent authorities,
- they must produce waste disposal plans,
- any installations or undertakings treating, storing or tipping waste on behalf of third parties must have a permit,
- the Polluter Pays Principle is to apply,
- Member States are to encourage recycling.

The plans must cover: the type and quantity of waste to be disposed of; general technical requirements; suitable disposal sites; and any special arrangements for particular wastes. The permits must cover: the type and quantity of waste to be treated; general technical requirements; precautions to be taken; and the information to be made available at the request of the competent authority.

## Netherlands

### FORMAL COMPLIANCE

Under the Wastes Act 1977 the licensing of waste facilities for domestic and industrial waste (apart from toxic waste) is the responsibility of the province which must also draw up a waste disposal plan. Under the Chemical Wastes Act 1976 the licensing of toxic waste is the responsibility of the Minister.

The Wastes Act 1977 and the Chemical Wastes Act 1976 did not come into effect until after the due date so that infringement proceedings were initiated against Netherlands for non-compliance. Even now there is still no requirement to draw up plans for the disposal of toxic waste so there could be a failure of formal compliance (see 5.2 Toxic Waste).

## EFFECT ON PRACTICE

No significant effects can be attributed to the Directive since Dutch legislation was being reformed before the Directive and the proposals for reform included the key provisions of the Directive. No full situation report has yet been submitted to the Commission.

# France

## FORMAL COMPLIANCE

The law on wastes of 1975 (which appeared simultaneously with the Directive) places the legal responsibility for collection and disposal of household waste on the *communes* but a circular requires working groups at the level of the *département* to prepare plans. There is no obligation to prepare plans for industrial waste but only a recommendation in a circular of 26 June 1980 to the Prefects of *régions* to set up regional working parties. There is therefore a failure of formal compliance.

## EFFECT ON PRACTICE

The establishment of regional working parties to plan for industrial waste can probably be attributed to the Directive which may also have applied pressure for a circular of 1983 requiring all unauthorised waste disposal sites either to be authorised or to be closed and their contents transferred to authorised sites.

No full situation report has yet been submitted to the Commission.

# Germany

## FORMAL COMPLIANCE

The federal law on the elimination of waste of 1972 was one of the inspirations for the Directive which in its turn inspired an amendment to the German law in 1977. The obligations of the Directive are to be found in the German law although the requirement about the content of plans is less defined than required by the Directive. The designation of competent authority varies between the *Laender*. There is no legal requirement in Germany that waste should be prevented and that recycling should be encouraged but there was a Federal Government 'programme' in 1975. There is doubt that the Polluter Pays Principle is fully applied since in some *Laender* (e.g. Bavaria) waste plants are subsidised.

## EFFECT ON PRACTICE

The Directive has had little effect on practice since it was modelled to a large extent on the German law on waste in its 1972 version. It may be influencing new legislation on prevention of waste and recycling. A brief situation report has been submitted to the Commission primarily describing the legal position.

# United Kingdom

## FORMAL COMPLIANCE

All the elements of the Directive are to be found in Part I of the Control of Pollution Act 1974. However the Act was not brought into force until 1978 so that formal compliance was one year late. Undoubtedly the Directive created pressure for the relevant section of the Act to be introduced.

## EFFECT ON PRACTICE

The Directive has not had any noticeable effect on practice apart from the pressure to bring into force the requirement to produce plans.

# General assessment

In all four countries legislation or draft legislation was in existence before the Directive was proposed. Interchange of ideas between the draft Directive and the draft national legislation may be assumed, but nevertheless there were delays over formal compliance in Netherlands, France and Britain and it is not complete in all the four countries. Thus there is no legal requirement in France that plans must be produced for industrial waste and no legal requirement in Netherlands that plans must be produced for toxic waste. There is doubt about the application of the Polluter Pays Principle in Germany. France and Netherlands have still to submit formal situation reports to the Commission although the Commission appointed consultants to review the position in all Member States so that some information has been made available to the Commission through these consultants.

## 5.2 TOXIC WASTE — DIRECTIVE 78/319
Date for formal implementation — 22.3.80

## Brief summary of Directive

The Directive is a complicated one which repeats and adds to the provisions of the framework Directive on waste (5.1). The following are the eight most important elements:

- toxic and dangerous waste is defined albeit rather loosely in some respects,
- general duties are placed on Member States to ensure that toxic waste is disposed of without harming the environment, and to encourage prevention and reuse of toxic waste
- competent authorities must draw up plans
- establishments storing, treating or depositing toxic waste (including establishments handling their own waste) must have a permit
- undertakings producing, holding or disposing of toxic waste must keep records
- an identification form must accompany toxic waste when transported
- the Polluter Pays Principle is to apply
- situation reports are to be submitted every three years.

## Netherlands

### FORMAL COMPLIANCE

A Decree of 1977 made under the Chemical Wastes Act 1976 defines chemical waste as material which contains any of 83 listed substances in greater than specified concentrations together with waste from designated industries. The Minister is himself the competent authority for authorisation but there is no legal obligation on him to prepare a plan. This could be a failure of formal compliance although in practice an indicative multi-year programme is drawn up and could constitute the plan if a broad policy statement can be considered as a 'plan'. There are a number of other failures of compliance: (a) the storage and treatment by an undertaking of its own toxic waste does not require a licence, (b) there is no provision for an identification form to accompany toxic waste when transported, (c) undertakings producing toxic waste do not have to keep records although the licence to transfer or dispose of toxic waste should ensure that a record exists.

## EFFECT ON PRACTICE

Since several features of the Directive are not formally complied with, it is hard to say that the Directive is having much practical effect, though it must create a pressure for changes to Dutch legislation. The definition of 'chemical waste' evolved with the Directive and was probably influenced by it. No situation report has yet been submitted to the Commission.

# France

## FORMAL COMPLIANCE

Waste is regarded as toxic if it includes substances listed in the 1977 *Décret d'application* of the 1975 law on wastes and there is no reference to specific concentrations of substances regarded as dangerous.

Toxic waste establishments are authorised at the level of the *département* and facilities recovering waste from *Installations Classées* are themselves *Installations Classées* and thus subject to special authorisations procedures. A circular of 26 June 1980 *recommended* the formation of working groups at the level of the *région* to draw up regional plans for industrial waste. The lack of an obligation to draw up these plans constitutes a failure of formal compliance. Producers, importers and transporters of toxic waste have to furnish information at the request of the authorities. The transport of dangerous substances was controlled by *le règlement sur le transport des matières dangereuse* (RTMD) of 15 April 1945. The RTMD runs to 700 pages, is complicated, and there was doubt whether it adequately required an indentification form to accompany toxic waste when transported. As a result the Commission issued a 'reasoned opinion' to the effect that the Directive was not fully implemented. In January 1985 a new *arrêté* required an indentification form in accordance with the Directive so that full implementation was over four years late.

## EFFECT ON PRACTICE

There have been considerable improvements in the system of handling toxic waste in France but these were being put into effect even without the Directive. The Directive is however probably responsible for the circular of 1980 recommending toxic waste plans at the level of the *region* and is responsible for the *arrêté* making obligatory an identification form to accompany toxic waste during transport. The list of toxic wastes evolved with the Directive and may have been influenced by it. No formal situation report has yet been submitted to the Commission but information has been given to consultants appointed by the Commission.

# Germany

### FORMAL COMPLIANCE

The definition of 'special wastes' under the Federal law of 1977 and a decree of 1977 is by reference to forms of production which produce named substances and is therefore more precise and probably narrower than in the Directive since in theory the substances could arise in waste from other sources. A decree of 1978 requires identification forms to accompany waste being transported. Other elements of the Directive are in the law of 1977 including the keeping of records for the production of waste but not the general obligation to promote prevention and reuse of toxic waste although these objectives are mentioned in Federal programmes and in those of the *Laender*.

### EFFECT ON PRACTICE

The definition of 'special wastes' evolved with the Directive and may have been influenced by it. The effect of the Directive has been very limited since it was to a large extent inspired by existing German legislation. The Directive has however served to underline the German legislation and ensures that it is taken seriously. It may also be influencing the new law on prevention of waste and recycling. A brief situation report has been submitted to the Commission primarily describing the legal position.

# United Kingdom

### FORMAL COMPLIANCE

The definition of 'special waste' that appears in Regulations of 1980 made under the Control of Pollution Act 1974 is by reference to the concentration that can harm human health if ingested and not by effects on the environment and so is a little narrower than in the Directive.

All the elements of the Directive are to be found in the Control of Pollution Act 1974 and other legislation except for two omissions which constitute failures of formal compliance: (a) undertakings producing or storing toxic waste do not have to keep records and (b) although identification forms must accompany toxic waste when transported, this does not apply when the toxic waste is being transported for recycling as also required by the Directive.

**EFFECT ON PRACTICE**

The definition of 'special waste' evolved with the Directive and was partly influenced by it. The Directive created pressure for the early introduction of the Regulations which among other matters requires identification forms to accompany toxic waste when transported. Otherwise the influence of the Directive has been slight. A situation report has been submitted to the Commission.

# General assessment

In all four countries legislation with provisions for toxic waste existed before the Directive was agreed in 1978, but in each country it was still evolving while the proposal was under discussion from 1976 onwards. Thus the national definitions of toxic waste were made in 1977 in Netherlands, France and Germany and in 1980 in UK. There was interaction between the national definitions and the definition in the Directive, but it remains a problem that the definition in the Directive is loose and that there is not yet any common definition throughout the Community. The problem arises when toxic waste is moved from one country to another and has not been resolved by the new Directive 84/631 on trans-frontier shipment of such waste. In Netherlands and UK the definition is by reference to the concentration of listed toxic substances, while the French and German definitions do not refer to concentrations.

In Netherlands, France and Britain there are failures of compliance. Thus there is no provision for identification forms to accompany toxic waste during transport in Netherlands. The keeping of records for the production of toxic waste is not obligatory in Netherlands and Britain, and there is no obligation in the Netherlands for a permit for establishments storing and treating their own waste. There is no obligation for a toxic waste disposal plan in Netherlands and France.

One of the main differences in administration between the four countries is that in Germany and Britain the same competent authorities are responsible for domestic and toxic waste and a single plan can cover both, whereas in France and Netherlands separate arrangements apply to toxic waste.

France and Netherlands have still to submit formal situation reports to the Commission although the Commission appointed consultants to review the position in all Member States so that some information has been made available to the Commission.

## 5.3  DISPOSAL OF PCBs — DIRECTIVE 76/403
### Date for formal implementation — 9.4.78

## Brief summary of the Directive

The Directive is rather general and lacking in precise provisions. Following a definition of 'PCB' and of 'disposal' four duties are placed on the Member States:

- to prohibit the uncontrolled discharge or dumping of PCB,
- to make compulsory the disposal of waste PCB and PCB contained in equipment no longer capable of being reused
- to ensure that PCB is disposed of without endagering human health and without harming the environment
- to ensure, as far as possible, the promotion of regeneration of waste PCB.

To carry out these duties the competent authorities are to set up or designate the undertakings authorised to dispose of PCB. The Polluter Pays Principle is to apply. Situation reports are to be drawn up every three years.

## Netherlands

### FORMAL COMPLIANCE

PCB is subject to the general provisions of the Chemical Wastes Act of 1976 and its disposal is regulated in the same way as other toxic waste. However discarded objects (e.g. transformers) were exempted until a Decree of 30 December 1983 removed the exemption for those objects containing PCB. For a time there was therefore a failure of compliance.

### EFFECT ON PRACTICE

Because there is no suitable high temperature incinerator in the

Netherlands, most waste PCB has been exported to Britain or Germany for incineration. Three facilities are however licensed to receive waste PCB but because the 1983 Decree is so recent only small quantities have been notified. The Directive will have created pressure for the 1983 Decree. A recent scheme providing a subisidy to encourage replacement of equipment containing PCB probably infringes the Polluter Pays Principle in the Directive. No situation report has been submitted to the Commission.

# France

## FORMAL COMPLIANCE

PCB is subject to the general provisions of the framework law on the elimination of waste of 15 July 1975, and also to an inter-ministerial *arrêté* of 8 July 1975. This *arrêté* is specific to PCB and contains the elements of the Directive.

## EFFECT ON PRACTICE

The French legislation preceded the Directive and indeed provided the stimulus for it. The Directive cannot therefore be said to have had much effect in France. Notwithstanding this early legislation, it does not seem to have been well applied since, as a result of the search for the lost drums of dioxin from Seveso in 1983, large quantities of PCB were found in waste disposal sites. The flows of PCB are not well known. No situation report has been submitted to the Commission.

# Germany

## FORMAL COMPLIANCE

There is no specific legislation for disposal of PCB. However products containing PCB are covered by the law on waste of 1977, particularly the provisions dealing with special waste. But there is some confusion over the law on elimination of waste oils because the definition of waste oils includes PCB. Since PCB should be eliminated separately from waste oils there is doubt whether the law thus ensures that PCB is eliminated without danger and so some doubt about formal compliance.

## EFFECT ON PRACTICE

It is doubtful whether the Directive, being so general, has had any effect on practice. There have been problems with the disposal of PCB in Germany. A saltmine is authorised to receive PCB and there are four authorised incinerators. The TA luft (technical instruction on air pollution) requires the burning of PCB in incinerators to be at 1200°C. Accidents have shown that PCB is sometimes attempted at inappropriate sites. A proposal to reform the law envisages that PCB disposal would be controlled only by the law on waste. Although a situation report has been submitted to the Commission it is limited to legal aspects.

# United Kingdom

### FORMAL COMPLIANCE

There is no specific legislation on disposal of PCB. PCB disposal is however subject to Part II of the Control of Pollution Act 1974 concerned generally with waste disposal, and to air pollution legislation. Regulations concerned with toxic waste include PCB and any transport of waste containing at least one per cent by weight of PCB has to be notified in advance and must be accompanied by a consignment note. There is however no legislation to make compulsory the disposal of waste PCB, e.g. in a disused transformer, so in this respect there is a failure of formal compliance.

### EFFECT ON PRACTICE

A 'Waste Management Paper' on PCB disposal preceded the Directive and it is unlikely that the Directive has had any effect on practice. There has been controversy over the incineration of PCB. A situation report has been submitted to the Commission.

# General assessment

The Directive is written in very general terms and contains some ambiguities. Because of this there are doubts about some aspects of formal compliance. In the Netherlands the exemption until 1983 of transformers from the Chemical Wastes Act constituted a temporary failure of compliance, and the recent subsidy for replacement of equipment containing PCB may well constitute a breach of the Polluter Pays Principle that is restated in the Directive. In France there is a requirement for disposal of disused transformers containing PCB as required by the Directive but not in the other three countries. The disposal of small capacitors containing PCB with household waste is accepted practice in all four countries. There is no incinerator for PCB in Netherlands. Although no situation reports have been submitted to the Commission for France and Netherlands, and the German report is confined to legal matters, some information has been made available to consultants appointed by the Commission.

# 5.4 WASTE OILS — DIRECTIVE 75/439
## Date for formal implementation — 18.6.77

## Brief summary of the Directive

A general duty is placed on Member States to ensure the safe collection and disposal of waste oils and that they are 'as far as possible' recycled. Prohibitions are placed on disposal to water and to soil, the uncontrolled discharge of residues from processing, and any processing causing air pollution. An optional system is provided for under which collection undertakings organised on a zonal basis must collect oil in the zone. Subsidies may be paid to these undertakings. Undertakings disposing of waste oils must all have a permit. Holders of contaminated waste oils must handle and stock them separately. Establishments producing, collecting or disposing of more than a certain quantity must keep records. Situation reports are to be submitted to the Commission.

## Netherlands

### FORMAL COMPLIANCE

The Chemical Wastes Act of 1976 specifically covers waste oils, but they are defined more narrowly than in the Directive. There are further failures of formal compliance in that there is no requirement that holders of contaminated waste oil must handle and stock them separately, and there is no general prohibition on the discharge of waste oils to sewers.

### EFFECT ON PRACTICE

The Chemical Wastes Bill preceded the Directive and provided the main stimulus to it, so it cannot be said that the Directive has had much effect in the Netherlands. Part 3 of the Chemical Wastes Act provides for the establishment of an organised network of undertakings for the collection and disposal of waste oil. Thus the optional system of the Directive has been put into effect with 40 collection areas in each of which at least one licensed facility is obliged to accept any oil in quantities of 400 litres or more offered to it. Most waste oil collected is burnt in small heating installations and recycling is insignificant in the Netherlands. No situation report has been submitted to the Commission.

# France

## FORMAL COMPLIANCE

General waste legislation applies to waste oils but following the second oil price rise in 1979 a decree was introduced obliging holders of waste oils to pass them to an authorised collector or to transport them themselves to an undertaking authorised to dispose of waste oils. There is an authorised collector in each *département* who must collect stocks of more than 200 litres and is in a monopoly position. From 1979 to 1981 a para-fiscal tax was placed on new oil in order to finance investment in recycling facilities and subsidies can be paid for storage, collection and treatment as allowed by the Directive. There is other legislation concerned with e.g. separate storage of waste oils.

## EFFECT ON PRACTICE

The post 1979 measures were strongly influenced by the Directive. However in practice there have been many problems and the comprehensive regulations have not been wholly effective. France is the only Member State of the EC which exports waste oils largely because of the low prices in France but also because laxer standards for burning waste oils are applied in other countries (e.g. Belgium). The monopoly position of the authorised collectors has been breached unofficially and France has been brought before the European Court by the Commission for regulating against the export of waste oils. As a result the French government is reviewing its legislation. There is a feeling in France that it is disadvantaged because it applies the Directive more rigorously than other Member States. No formal situation report has been submitted to the Commission.

# Germany

## FORMAL COMPLIANCE

The law of 1968 on waste oils which dealt with collection on a zonal basis and the payment of subsidies (and which had probably inspired the Dutch legislation) had to be amended in 1976 and again in 1979 before formal compliance was assured. The 1979 amendment dealt with separate storage for contaminated oil so that there was a delay over formal compliance. Prohibition of discharges to water and to soil is dealt with by general water and waste legislation. The subsidies are financed by a levy on new oil.

## EFFECT ON PRACTICE

The Directive has influenced German practice only on points of detail, e.g. separate storage, since the German system was well developed before the Directive. The subsidies are being progressively eliminated and the view has been expressed that this could pose problems in ensuring that waste oils are disposed of without harm to the environment.

# United Kingdom

## FORMAL COMPLIANCE

There is no legislation specifically dealing with the disposal of waste oils. Waste oils are however subject to environmental legislation dealing with discharges to water and air and deposits to land. There is no requirement that contaminated oil should be stored separately and no requirement that producers and collectors of waste oil should keep records. There is therefore a failure of formal compliance. The optional provision for subsidies to be paid to collectors authorised on a zonal basis has not been used in Britain.

## EFFECT ON PRACTICE

The Directive has not had any effect on practice in Britain. A situation report has been submitted to the Commission.

# General assessment

The Directive has had a significant effect in France but little effect in the other three countries. The greatest difference in implementation of the Directive is that in Britain, unlike the other three countries, there are no subsidised waste oil collection and disposal undertakings specially authorised on a zonal basis. This is an optional provision of the Directive so this difference does not means a failure of compliance in Britain. In Britain the oil recovery industry, as represented by the Chemical Recovery

Association, believes that the intense competition among a large number of small recovery firms ensures that all recoverable oil is indeed recovered and therefore there is no need to introduce the optional system of the Directive. The rather rigid French system is being reviewed following a legal challenge from the Commission.

There is a failure of formal compliance both in Netherlands and in Britain in that holders of contaminated oil do not have to handle and stock them separately. There is a further failure in Britain in that there is no requirement that producers and collectors of waste oil must keep records.

# 6

# Formal Implementation

According to Article 189 of the Treaty of Rome a Directive is not, unlike a Regulation, directly applicable in the Member States but merely specifies the result to be achieved leaving 'to the national authorities the choice of form and methods'. The Commission therefore does not exercise control over the choice of form and methods: what matters is the result — even if in some respects certain Directives are so precise that they leave the Member States little latitude.

If one analyses the results that are to be achieved by the environmental Directives, one can say that the Community imposes four different kinds of obligation on the Member States:

- the need to adapt national legislation, administrative structures and procedures so that they conform to the rules and procedures set out in the Directive
- the need to put these rules and procedures into practice
- the need to ensure that the quality of the environment meets the standards or the ends set out in the Directive
- the fourth obligation concerns communications that the Member States must make to the Commission to enable the Commission to exercise control over the other three obligations.

There are therefore two categories of failure to comply with Directives, or in other words to implement them. First, those which can be called *formal* and concern legal relations in the Community, including the primacy of Community law over national law which must conform to it, and the obligation of national authorities to give an account to the Community institutions. Second, those which concern *practical* implementation of Community rules and which require the achievement of practical results (quality of water, the putting in place of authorisation procedures, monitoring, designations, drawing up plans and programmes, etc).

This chapter discusses only *formal* implementation and does so under the two headings below.

# 1. THE OBLIGATION TO COMMUNICATE

## 1.1 National legislative and administrative measures

Chronologically, the first obligation is for the Member States to communicate to the Commission the national laws and administrative measures on which they are relying to enable them to apply the Directive. There is usually a specified time period (often two years) to satisfy this obligation. All Member States have made communications but there have been several problems.

● *Delays*. All Member States have on occasion been late with their communications, and the Commission has frequently had to remind Member States of their obligations. In numerous cases, the Commission has used the procedure under Article 169 of the Treaty. This procedure involves three phases. First a letter is sent starting the procedure and telling the Member States that the Commission is examining the matter to see if there is an infringement. Second a 'reasoned opinion' is issued under which the Member States are asked to put the situation right within a prescribed period. Third, there will be a request to appear at the Court of Justice. In effect there does not necessarily have to have been a substantive failure of compliance since the Member State may well have taken the appropriate measures but may have failed to communicate them to the Commission. This has happened in Germany, for example. Given the legislative competence of the *Laender* in the field of water, the Federal Government can only transmit to the Commission what the 11 *Laender* have themselves transmitted. These two stages of transmission are undoubtedly one of the reasons for the numerous delays in the communications from Germany. The same problem has sometimes arisen in the United Kingdom where different legislative and administrative measures may have been applied in Scotland and Northern Ireland from those in England and Wales. In numerous cases there have to be several communications from the Member State to the Commission because earlier communications have had to be supplemented.

● *The content of these communications is very variable*. Sometimes the Member States merely quote the national measures taken (law, regulations, circulars, etc.) although more usually they send the full text to the Commission which may otherwise ask for them. Occasionally the Member States send an analysis indicating for each Article of a Directive the precise national measures that implement it.

• *The method of transmission.* All the Member States maintain Permanent Representations in Brussels which are the equivalent of embassies to the Community institutions. The responsible Ministries in the national capitals will draft a so-called 'compliance letter' setting out how an individual Directive is being complied with, but it will be the Permanent Representation that will send the letter to the Commission. These 'compliance letters' are not in general made public but they nevertheless represent an important link between Community legislation and national legislation. The British 'compliance letters' relating to the water and waste Directives were made available for the British national report on which this comparative report is based, and the Federal Republic of Germany has declared itself ready to do so for the future. The other two countries have yet to do so.

## 1.2 Situation reports on specific aspects of the environment

Several Directives require the Member States to submit reports to the Commission on specific aspects of the environment covered by the Directive in question. These may relate to the standards set in the Directive, or to pollution reduction programmes required, or to lists of waters designated for a particular purpose, etc. Sometimes the Commission is in turn required to consolidate the national reports and sometimes to publish the results. There have been numerous delays with these reports, some of which must amount to a formal failure to comply with a Directive. The manner in which the material in the national reports is presented frequently varies considerably and creates problems for purposes of comparison. An example is provided by the national reports on bathing water quality which vary from a thick volume for France to only a few pages each for Germany and the United Kingdom, and which also vary in the technical material presented.

## 2. ADAPTING NATIONAL LEGISLATIVE AND ADMINISTRATIVE MEASURES TO CONFORM WITH DIRECTIVES

The practice in the different Member States shows that several approaches have been used including the introduction of a new law in parliament, regulations made under existing laws, simple administrative circulars, and even reliance on the status quo. Besides the choice of instrument there is the question of conformity itself, that is to say the extent to which Community obligations have indeed been transcribed into national measures and the further question of the control of this conformity.

In all the national legal systems the most binding instrument is a law or Act of the parliament. A law will usually state the principles and essential rules but may also authorise the government to specify the details of these rules so that they can be put into effect by the responsible authorities and those on whom obligations are placed. It will itself state, or give the government the power to state, which authorities are to be responsible. The form of the instrument by which a government exercises the power given it by a law varies, e.g. decree, *arrêté*, regulation, statutory instrument, command. If the administrative authorities already possess the necessary powers to achieve the results set out in a Directive it is only necessary to specify these results and to give advice where the obligations are not precise, and this can be done by a simple administrative circular.

## 2.1 Instruments

How, bearing in mind this scheme, have the water and waste Directives acquired a mandatory character under national law? Let us put aside the case of parallel development of a Directive and a national law corresponding to the same subject, where one can assume, without always being able to demonstrate the point, that there have been reciprocal influences. The French law on wastes of 1975 and Directive 75/442 (Section 5.1) is one case, and the German water law of 1976 and Directive 76/464 (Section 3.8) is another. Some countries have laws giving very general powers to implement Community obligations. In the United Kingdom the European Communities Act 1972 provides such powers but practice shows that there is a preference for using other legislation unless it is insufficient for putting the Directive into effect. The detergent Directives (Section 3.1) were implemented using these powers. In Germany a very general provision was introduced in 1976 into the law on water under which water management plans must be drawn up 'if it is necessary to fulfil Community obligations'. Certain *Laender* have even introduced general references to Community legislation in their own laws on water, to allow the authorities to refuse discharge authorisations if Community rules make it necessary. In the Netherlands the Surface Waters Pollution Act 1969 was amended in 1981 so that by a Ministerial decree the provisions of international agreements, including EC Directives, can be implemented.

The amendment in 1981 to the Surface Waters Act 1969 in the Netherlands is the most striking case of the introduction of new legislation to implement Directives considered in this study. It followed a judgement of the Court of Justice concerning failure to implement Directives 75/440 and 76/160 (bathing water and surface water — Sections 3.7 and 3.2) but has enabled other water Directives to be implemented as well. New legislation is also being introduced in the Netherlands in order to implement Directive 80/68 on groundwater.

Government regulatory instruments have been used in several cases to specify obligations in the Directive. For example: in the Netherlands decrees have been made setting out quality objectives for four types of water, fixing emission standards and specifying values for parameters for drinking water; in Germany the Federal Government is preparing a decree on the quality of drinking water: in France an *arrêté* has been used to specify the quality of bathing water; and in the United Kingdom Regulations have been made defining toxic waste and requiring an identification form to accompany toxic waste when being transported.

In three countries — the Netherlands being the exception — the majority of Directives have been implemented by government circulars addressed to the relevant authorities. This can be done if legislative measures are already in existence covering the obligations in a Directive and also if the central government has the means to ensure that the obligations set out in the Directive will be carried out. Such was the case with nearly all the Directives in the United Kingdom and for the majority in France and in the German *Laender*. Circulars have two roles: they draw the Directive to the attention of the authorities and indicate to them how they are to interpret national legislation to take the Directive into account. Sometimes there may be circular if the Member State considers that a Directive closely corresponds to previously existing legislation. This was the case in the United Kingdom with Directive 75/442 on waste which is very similar to Part I of the Control of Pollution Act 1974.

Implementation by means of circulars has not caused particular difficulties so far, but two problems can be foreseen. First, not all circulars are widely publicised, since they are not always officially printed, perhaps in the national Official Journal, with the result that there is no publicly visible link between Community legislation and national legislation. Lack of publication could have consequences when a Directive indirectly creates rights and obligations for individuals since it is desirable that those affected should know the source of the obligations imposed on them. Second, failure to publish can result in a lack of control over the authorities whose duty it is to implement Directives since public ignorance will mean that there will be no public pressure.

# 2.2 Conformity

There have been delays in formally implementing all the Directives considered here. Most have now been formally implemented in all four countries although some failures still remain. Either as a result of these delays or because of doubts over the adequacy of existing national measures, the Commission has on numerous occasions begun infringement procedures under Article 169 against the Member States. Although several 'reasoned opinions' have been issued, only two cases have come before the Court of Justice which found in favour of the Commission. As mentioned above these cases involved the Netherlands and concerned Directive 75/440 on surface water (3.2) and Directive 76/160 on bathing water (3.7).

Any discussion of failure of formal implementation needs to be surrounded by a number of qualifications since not all failures are equally serious and not all are equally culpable. Often a failure of formal implementation is only a procedural failure since the practical obligations in the Directive may have been carried out anyway. An example of this is the detergent Directives (Section 3.1) where legislation was late in all four countries but where at least in the Netherlands and the United Kingdom the standards were already being observed by the manufacturers on a voluntary basis. It also sometimes happens that the greater part of a Directive will be fully implemented while some relatively minor part may be delayed. There may also be genuine doubt about whether formal implementation is complete since the internal logic of each national legal system means that precise equivalences often do not exist and a judgement then has to be made as to whether a national provision is or is not equivalent to a Community obligation. This is particularly the case when an obligation has been expressed in rather general wording in a Directive, either because it has been poorly drafted or because of the need to secure political agreement by all Member States. The national measures implementing the rather general wording may nevertheless have to be precise in order to be workable. An example is provided by Directive 78/319 on toxic waste (Section 5.2) which contains a definition of toxic waste in rather loose terms. The four Member States when making their own definitions have had to conform to the Community definition but the national definitions all differ from one another and are in places more precise than that in the Directive and so are arguably narrower.

A particularly interesting case of doubt about formal implementation has arisen with Directive 78/659 on waters for freshwater fish (Section 3.5) and Directive 79/923 on shellfish waters (Section 3.6). In Germany no waters have been designated under either Directive by the *Laender*.

Clearly there has been no practical implementation of the Directive but has there been a failure of formal implementation? Since the Directive leaves it to the Member States themselves to make the designations there is in effect no obligation to do so. However no legislative or administrative steps have been taken in Germany to implement the Directives. This could be regarded as a failure of formal implementation, but on the other hand it can be argued that since designation is discretionary no administrative measures are needed once the decision to make no designations has been taken.

The failure by both France and the Netherlands to submit reports to the Commission concerned with the situation on waste disposal as required by all four Directives on waste can also be regarded as a failure of formal implementation. However, both countries have supplied information to consultants employed by the Commission and this could possibly be held to satisfy the requirements of the Directive.

The fact that formal implementation has caused difficulties in all four countries can be seen from the often lengthy delays that there have been. In the Netherlands the long delays in implementing several of the water Directives — and that led to adverse judgements by the Court of Justice — resulted from the need to introduce new legislation, legislation which has itself significantly changed the system of water management in the Netherlands. The slow and meticulous nature of the Dutch legislative process has been one cause of the delay. There have also been delays in completing the implementation of several water Directives in the United Kingdom. Although the necessary powers existed in Part II of the Control of Pollution Act 1974, they have only been brought into force recently. The delay here has been caused by the desire of the Government to defer increases in public expenditure and to hold down costs to industry, and the existence of the Directives has undoubtedly created a pressure for finally bringing the Act fully into force. In France there have also been delays including a delay of three years in the case of the bathing water Directive 76/160 (Section 3.7) and over four years in the case of the toxic waste Directive 78/319 (Section 5.2). There has been a delay in Germany over the drinking water Directive 80/778 (Section 3.4).

The fact that delays have been frequent illustrates one of the difficulties in policy making at a Community level. The process of negotiation in the Council is to some extent an attempt by the Member States to ensure that a Directive is agreed in such a form that it is compatible with existing national legislative and administrative practices. Inevitably compromises have to be made and countries may well find themselves accepting commitments before they are ready for them. Some delays are to be expected and should be tolerated if they are not too frequent or too extended. If they become too frequent or too extended then Community legislation is brought into disrespect. The question must nevertheless be asked as to whether, in the negotiations leading to the adoption of a Directive, countries admit to themselves and to others that they know they cannot meet the date for implementation. Since negotiations take place largely behind closed doors it is a question that outsiders can only ask without expecting an answer. Countries who are consistently late cannot however expect to be taken seriously in the future when they insist upon urgency in agreeing new Community proposals.

For the reasons given above a catalogue of failures of formal implementation runs the risk of giving a misleading impression. Explanations are nearly always forthcoming but are not always easy to assess. Whatever the reasons it is possible to say that in all four countries there are still some definite failures of formal implementation of the Directives considered here and some where a failure is possible. Some of these are set out below. It has to be emphasised that this catalogue is likely to become out of date as Member States adopt new legislative and administrative measures.

# Netherlands

| | |
|---|---|
| Dangerous substances (3.8) | No powers for Government to ensure that quality objectives are set for List II substances |
| Groundwater (3.9) | New Ground Protection Act required |
| Waste (5.1) | Doubt whether the obligation for a waste disposal plan is fulfilled so far as toxic waste is concerned |
| Toxic waste (5.9) | Same point as in (5.1) above. Also no obligation for a licence to treat and store own toxic waste. Also no requirement for an indentification form to accompany toxic waste |
| Waste oils (5.4) | No requirement that holders of contaminated waste oils must store them separately |

# France

| | |
|---|---|
| Drinking water (3.4) | New decree required to fix MAC values |
| Groundwater (3.9) | Doubt whether existing provisions are adequate. New circular in preparation |
| Waste (5.1) | No obligation to prepare disposal plans for industrial waste |
| Toxic waste (5.2) | No obligation to prepare disposal plans for industrial waste |

# Germany

| | |
|---|---|
| Drinking water (3.4) | New draft decree being prepared |
| Freshwater fish (3.5) ⎫<br>Shellfish waters (3.6) ⎭ | Possible doubt over formal implementation since no waters have been designated (see above) |
| Disposal of PCBs (5.3) | Doubt created by overlap with the law on waste oils |

# United Kingdom

| | |
|---|---|
| Toxic waste (5.2) | No requirement for records of production or for identification forms to accompany waste intended for recycling |
| Disposal of PCBs (5.3) | No requirement for the compulsory disposal of waste PCBs, e.g. in an unused transformer |
| Waste oils (5.4) | No requirement for records for establishments producing and collecting waste oils. No requirement for contaminated oil to be kept separately |

# 7

# Practical Implementation

The problems surrounding the implementation of public policy have become the focus of an increasing number of studies in recent years. As a result, the difficulties commonly besetting the implementation process — underachievement of stated objectives, delay and excessive cost — are now widely appreciated. Less well understood, however, are the factors which operate to hinder effective implementation. What appears to occur is that policy implementation in a national context is as much a political activity as policy formulation. That is to say, the process of policy implementation — just like policy formulation — is characterised by an array of conflicting interests (the enforcement agency, other government departments, private companies, interest groups) each seeking to achieve its own objectives and resorting to persuasion and bargaining to do so. Resistance to a particular policy proposal does not simply disappear with the conclusion of the legislative process.

It is to be expected that the implementation of EC Directives will give rise to even greater difficulties than the implementation of purely national legislation. As explained in Chapter 1, not only is the chain between Community legislation and action on the ground one link longer, but the mechanics of enforcement is a matter for the Member States themselves and competent authorities within them who will not have had much say in the drafting of the Community legislation. In Chapter 6 we have pointed to the often long delays that there have been before national measures are introduced formally to implement the Directives. The result is that practical implementation is also delayed, so much so that any judgement on the impact of the Directives must still be provisional.

A further factor that complicates any judgement is the view expressed in several countries by the officials responsible, when they were interviewed in the course of the four national studies, that they would anyway have done what was required by a Directive. Such statements are very difficult

to prove or disprove. To counterbalance this many admitted that
Directives had on occasion provided the impulse for change particularly
when arguments were needed to justify expenditure. Directives have also
often strengthened the hands of the authorities when negotiating with
industrialists since they provide standards which are not negotiable —
standards which have a higher status than national standards since in
theory they are being applied in all Member States. Despite these many
difficulties in making judgements it is possible to point to a number of
practical changes that in all probability are the direct result of the
Directives. The following should not however be regarded as a complete
catalogue.

*Water*

| | |
|---|---|
| 3.1 Detergents | Improvements to the composition of detergents sold in France |
| 3.2 Surface water | Certain sources of poor quality drinking water have been abandoned in France |
| 3.3 Sampling surface water | More parameters being measured in all four countries |
| 3.4 Drinking water | Premature to judge full effects but already clear that important effects are being felt in all four countries. In particular attention has been focussed on the problem of nitrates |
| 3.5 Freshwater fish water | Substantial designations in UK but not necessarily leading to improvement of water quality. No designations yet elsewhere |
| 3.6 Shellfish waters | Some designations in UK but not necessarily leading to improvement of water quality. No designations yet elsewhere |
| 3.7 Bathing water | Rather different practices with respect to identification of bathing waters has been adopted in the four countries. No formal designations yet in Netherlands. Twenty five per cent of French designations do not meet standards so a pressure for improvement has been generated. More is understood now about the quality of bathing waters in all four countries |
| 3.8 Dangerous substances | Little effect on practice although a significant effect on policy (see Chapter 8). However attention has been focussed on List I and List II substances |
| 3.9 Groundwater | Little practical effect beyond a reinforcement of national policies in some countries |
| 3.10 Mercury | No effect in Netherlands and Germany. Pressure for improvements in France and UK |

| 3.11 | Titanium dioxide | Pressure for improvements in all four countries. The requirement for improvement plans has generated a dialogue between Germany and Netherlands about dumping at sea. |

*Waste*

| 5.1 | Waste | Little effect on practice beyond emphasising the need for waste disposal plans |
| 5.2 | Toxic waste | The national definitions of toxic waste were most probably influenced. The introduction in France of an identification form to accompany waste during transport is a result of the Directive. This is still not required in Netherlands |
| 5.3 | Disposal of PCBs | Being so general it is not clear that this Directive has had any practical effect |
| 5.4 | Waste oils | The collection system on a zonal basis in France was most probably influenced by the Directive. |

This catalogue of practical effects is offset by some significant failures fully to implement the Directives in practice. Many of the delays and failures over formal implementation noted in Chapter 6 also mean that there has not yet been any practical implementation. The delay in the Netherlands in designating bathing waters, shellfish waters and waters for freshwater fish constitute failures of practical implementation. In France shellfish waters and waters for freshwater fish are only beginning to be designated and in Germany a decision has been taken not to designate any waters for the time being. In the United Kingdom the principal failure of practical implementation is the consequence of the partial failure of formal implementation of the Directive on waste oils (5.4) — see Chapter 6. In all four countries the Directive on discharge of dangerous substances to water (3.8) is not yet fully implemented in practice since not all discharges of List I and List II substances are specifically authorised e.g. from sewage works.

What are the reasons for those failures of practical implementation which are not the simple consequence of delays in the introduction of national legislation? On the basis of this study two key features of Community policy formulation and implementation can be advanced to explain this.

First is the problem that stems from the contrasting perspectives on environmental management prevailing in the local and regional implementing authorities in the Member States from those prevailing among the policy-makers in the Community. While the Commission is concerned with the formation of a broad environmental policy applicable across the Community, the preoccupation of environmental control officers is the resolution of local problems. As such, the general principles

93

of control reflected in the Directives are not as a rule the primary concern of the officers in carrying out their day-to-day duties. Indeed, ignorance of the specific provisions laid down in the various Directives was not unusual. There was, of course, much greater familiarity with national legislation and policy objectives, but these are often structured in a very different way to the Directives and as a consequence may not convey the same 'intent' as a specific Directive. The remoteness of a control agency from Community policy making carries with it the risk that there will be no great commitment to Community objectives. The most obvious example is the opposition of the German *Laender* to the principle of designating waters in order to protect freshwater fish, and shellfish. Even where an agency looks beyond local problems it will first consider regional and national priorities — in practice 'basic quality' still constitutes the dominant thrust of water pollution control in the Netherlands, for instance. This problem can only be overcome by a much greater understanding among responsible officials of what happens in other countries and of the reasons for the existence of a Community environmental policy in the first place.

In implementing Community Directives a lack of commitment to the stated objectives will have far-reaching consequences. And this leads to the second source of implementation failure — the question of imprecise legislation and discretionary obligations. The bathing water Directive, for example, is characterised by a notably vague definition of what areas are to be regarded as bathing waters and Member States have exercised their discretion in identifying waters. The result in practice is that no two Member States apply the same criteria to the identification of bathing water, and in some cases the high cost involved in bringing poor quality waters up to the necessary standards has proved to be the key factor in a decision not to identify certain waters. Similarly the freshwater fish and shellfish waters Directives give Member States the complete discretion in designating waters to the extent that no designations are being made in Germany.

In a national context there are two approaches to the conflict between precision and discretion. Either the obligations are defined very precisely in legislation so that all the parties affected know where they stand, or a broad obligation may be laid down with the responsibility for exercising discretion being clearly given to a competent authority with the result that the authority will then have to negotiate with the bodies being regulated. It is perfectly obvious that Directives will frequently have to include indefinite language if political compromises are to be secured, but there is then the danger that the Directive will be applied in an inconsistent manner throughout the Community. The dilemma can be stated as follows: Directives will sometimes have to use indefinite language; there must be some consistency of application throughout the Community if a Directive is to achieve its purpose, and yet discretion must be applied by someone if worthwhile practical results are to be achieved. There is no simple way to resolve this dilemma but undoubtedly a greater knowledge among the competent authorities of how the same Directive is being applied in different countries could, in the long run, help to bring a greater measure of consistency. This comparactive report is a contribution to this process, and may in turn stimulate more communication between the competent authorities in different countries responsible for particular environmental Directives.

No comment has been made in this Chapter on the environmental improvements that are the whole purpose of implementing environmental Directives and that are indeed the reason for having them in the first place. Some of the changes in the above catalogue of practical effects involve improvements although some are just procedural. Nevertheless the procedural changes should lay the basis for improvements in the future. Above all Community legislation, as long as it does not fall into disrepute as a result of not being properly implemented, gives greater attention to environmental issues and makes it easier to justify the expenditure of money on which so many improvements depend.

# 8
# Impact on Policy

In addition to having practical effects — as discussed in Chapter 7 — the Directives considered here have also influenced national policies and can be expected to continue doing so as they become more thoroughly assimilated. A particular Directive does not cease to have an effect the moment it has been formally implemented, or the moment certain steps have been taken to achieve the practical results set out.

Although some policy changes have resulted from the waste Directives the effect of the water Directive has been greater very possibly because they embody theories of pollution control which have not been pursued in quite the same way in the four countries. In addition both the water and waste Directives have had effects on the relationship between central and local government and this unexpected and unintentional effect needs a brief discussion.

## 1. WATER

Taken together the water Directives embody two different approaches to water pollution control. On the one hand there is the functional approach to water management, and on the other hand there is the approach based on limiting discharges by reference to what is technically achievable. The functional approach involves indentifying different functions for different bodies of water, setting quality objectives for these functions, and then introducing measures to ensure that the quality objectives are achieved. By contrast the approach based on what is technically achievable concentrates on individual discharges without reference to the receiving waters.

If one leaves aside the four detergent Directives (Section 3.1) which set standards for products, one can say that the great majority of the water Directives embody the functional approach — an exception being the titanium dioxide Directive (Section 3.11). Even the Directive on the discharge of dangerous substances (Section 3.8) and its daughter Directive on mercury discharges (Section 3.10) involve a reference to the receiving waters even though the preferred approach for the List I substances is based on what is technically achievable.

Both these two approaches are to be found in all four countries but have developed in different ways and are applied with different emphases. As a result the water Directives have influenced policies in quite different ways.

In the **Netherlands** a new concept in water management was introduced just before the Directives come into effect. This is the concept of 'basic quality' — a range of parameters defining the minimum acceptable quality of all surface waters. This can be regarded as a special case of the functional approach with all 'functions' combined, although in the Netherlands it is regarded as a distinct approach in which function is not recognised. But the approach based on what is technically achievable is also in existence so that emission standards set for individual discharges following that approach also have to be such that the basic quality is achieved. However, since the Surface Waters Pollution Act of 1969 did not give the Minister power to require quality objectives to be set by reference to a number of functions of the kind set out in the Directives, the Act had to be amended to implement the Directives. It was also amended to give the Minister power to set emission standards at national level. The Directives have therefore resulted in a significant change in Dutch policy. It is a combination of this policy change and the need for new legislation that led to lengthy delays in formally implementing the Directives. Even now the process of practical implementation, including the designation of waters for fish, shellfish, and bathing, has only just begun.

In **France** both approaches existed before the Directives. The functional approach was introduced by the Water Law of 1964 and the approach based on what is technically achievable with emission standards prescribed at national level has long been applied to *Installations Classées*. Community concepts have therefore not influenced French concepts and indeed it was the other way about, with the functional approach set out in the law of 1964 influencing the Commission. Paradoxically, the functional approach has not been well developed in France before the Directives as is shown by the fact that only a small part of the country is yet covered by maps of water quality for different uses and by the fact that the designation of waters for fish and shellfish is only just beginning. The Directives are therefore having the effect of reinforcing the functional approach in France.

In **Germany** the influence of the Directives is rather more problematical since water policy was undergoing revision in the mid seventies at the time that the Directives were being introduced. The concept of regulating emissions by what is technically achievable did not exist in German law before 1976 but instead they were regulated by reference to the receiving waters (the 'immission' approach as opposed to the 'emission' approach — to use the German terminology). In moving to an approach based on what is technically achievable the *Laender* simultaneously moved away from the functional approach. This may be partly explained by the difficulty of applying the functional approach to such a long river as the Rhine coupled with the difficulty of applying it to rivers that cross administrative boundaries. There is therefore now a conflict between the functional approach embodied in the Directives and current German practice, a conflict sharply exemplified by the decision not to designate any waters for fish and shellfish, therefore rendering these Directives inoperative in Germany. Two other 'functional' Directives have been implemented, namely the Directive on surface water for drinking (Section 3.2) and bathing water (Section 3.7) although neither involves the designation of large areas of water. The integration of the functional approach into German practice will have to be a necessary task for the future if the functional approach embodied in the Directives is to be of a truly European scope.

In the **United Kingdom** the two approaches existed before the Directives but in the form of loosely applied concepts which were not formalised in legislation. When the Community proposed the Directive on the discharge of dangerous substances (Section 3.8) based on what is technially achievable the British saw this as contrary to their economic interests but presented their opposition to it also on theoretical grounds. The result was the compromise in the Directive that allows Member States to use an approach based on quality objectives as an alternative. Having opposed the approach based on what is technically achievable the British had to show their commitment to the quality objective approach and set about formalising it. Although it is possible that this development would have taken place anyway there can be little doubt that the Directives gave it an impulse.

# General assessment

This summary of the effect on policy shows the same Directives having quite different effects in different countries depending on the state of development of national policy. Perhaps the most striking comparison is between Britain and Germany where in both countries emission standards were set before the Directives in a rather unformalised way by reference to the receiving waters. The extent to which it was the Directives that actually caused the countries to move in opposite directions (Germany towards the approach based on what is technially achievable, Britain towards the functional approach) is a matter that deserves further attention. It is at least possible that these tendencies were under way but the Directives have helped to accentuate them. The fact that both approaches have existed in some form or other in all four countries suggests that both have a role to play. One of the tasks for the future development of the Community's water pollution policy is to relate the two approaches so that they better complement one another.

# 2. WASTE

By contrast the influence of the Directives on overall policy for waste in the four countries has not been dramatic and it is instead some points of detail that have been influenced.

In the **Netherlands** the Wastes Act 1977 and the Chemical Wastes Act 1976 developed in parallel with the Directives on waste (Section 5.1) and on toxic waste (Section 5.2) although several provisions of the Dutch legislation did not come into force till much later and some provisions still remain to be put into effect. The toxic waste Directive must, for example, have applied pressure for a national waste disposal plan for toxic waste and must be applying pressure for the requirement for identification forms to accompany toxic waste when being transported. The Directive on waste oils (Section 5.4) must be applying pressure for a requirement that contaminated oils be stored separately.

In **France** the proposal that became the 1975 Law on Waste developed at the same time as the waste Directive (Section 5.1) but nevertheless does not cover exactly the same ground. There is, for example, no requirement in the law for waste disposal plans although these were subsequently introduced in France by means of an administrative circular very probably in order to implement the Directive. The system introduced in January 1985 requiring toxic waste producers to follow their waste until disposal and to have their toxic waste accompanied by an identification form when being transported was also introduced as a result of the obligation in the toxic waste Directive (Section 5.2).

In neither **Germany** nor the **United Kingdom** have the waste Directives made any significant changes to policy. However the Directive on disposal of PCBs (Section 5.3) may be exerting an influence to clarify the legislation in Germany, and the Directive on waste oils (Section 5.4) could yet exert pressure for a requirement in the UK that contaminated oils be stored separately.

# General assessment

Possibly the most important broad provision of the waste Directives, which may not otherwise have been applied so thoroughly in all countries, is the requirement to prepare waste disposal plans. These plans are by no means yet complete but when eventually they come to be compared they could yet help to change policies as countries learn more about one another. Such comparisons could well, for example, show up different national policies for the landfill disposal of hazardous waste, incineration and recycling. A more positive note in Community waste policy has recently been struck by Directive 84/681 on trans-frontier shipment of hazardous waste, a problem that cannot be handled by nation states acting alone.

## 3. CENTRAL/LOCAL GOVERNMENT RELATIONSHIPS

One broad effect is common to both the water and waste Directives and is the consequence of the insertion of another layer of decision making on top of the existing layers at national and regional or local levels. The Treaty of Rome is such that obligations created by Community legislation fall on the Member States and it is they who are responsible for seeing that obligations are fulfilled even if the task is delegated to some sub-national authority or level of government. In several countries environmental matters, like many other matters, have in the past been largely local government functions. Since however the Community can only deal with central governments, what had previously been local functions have to become national government functions the moment they fall within a Community policy.

This process can be seen at work in the environmental field most clearly in the Netherlands. Until recently the management of certain waters was a provincial or local matter and central government had no power to set standards at national level that the provincial or other authorities had to apply. As we have seen, new legislation has had to be introduced in the Netherlands to fulfil Community obligations and standards are now being set at national level. The creation of a Community level of decision making has therefore been accompanied by a shift in decision making power from provincial/local level to national level. Paradoxically at the same time as national government lose some power — because of the inevitable need to make compromises in the Council of Ministers — they also have had to increase their powers over local authorities to ensure that these authorities fulfil Community obligations.

This process has taken place also in the United Kingdom even if it has not involved new legislation. Instead central government has controlled water authorities and local authorities by means of administrative circulars backed by the reserve powers over them that may previously not have been used. In France this shift of power is less noticeable because the authorities located at the level of the *département* responsible for practical implementation of the Directives are anyway the services of central government Ministries. The advice and instruction that these services receive from their Ministries is now however influenced by Community Directives. In practice one of the problems in implementation in France is that the services are not the services of the Ministry of the Environment but of other Ministries.

The shift of power from local to national level that is a consequence of Community policies raises particular problems in a federal state. The Basic Law in the Federal Republic of Germany apportions powers between the *Bund* and the *Laender* and with water, in particular, the *Laender* have competence within a framework federal law. The *Bund* is therefore dependent on the willingness of the *Laender* to implement the Directives and would not for example be able to compel them, even if it wanted to, to designate waters for shellfish or freshwater fish. If therefore the constitutional division of powers in Germany is not to be disturbed it is essential that the *Laender* are convinced of the need for the Directives before they are agreed, and this requires that the process of decision making in the Community is more open and more widely understood. Once Directives have been agreed, however, as is the case with the shellfish and freshwater fish Directives, quite a sharp problem is created if they are not properly applied in Germany. The Commission can take the Federal Government to the European Court which then creates a problem between the Federal Government and the *Laender*. There is no simple solution to this problem which goes well beyond environmental policy: it remains a factor that must be borne in mind when developing Community policies. There will be times when certain constitutional principles are of greater importance than the needs of a particular detail of a Community policy. On the other hand the *Laender* need to be aware that they may be preventing the development of a desirable Community policy. This awareness can best come about from knowledge of what happens in other countries.

# 9
# Summary and Conclusions

'Policy' can be regarded as the setting of certain goals and how they are to be realised. This report has been concerned with the implementation of EC water and waste policy in four countries, and thus with the extent to which the goals of these policies have been achieved and what effect they have had in the four countries.

Environmental policy in the EC has so far largely been pursued by the creation of legislation. Implementation of EC environmental legislation has been shown to have two distinct phases — *formal* compliance (i.e. the legal and administrative steps taken in response to a Directive) and *practical* compliance (i.e. what is done in practice to put the national measures into effect). In addition EC legislation can involve changes in ways of thinking about issues, and changes in the relations between central and local government. In other words EC policy can change national policy in often unexpected ways.

Because the starting point in each country is different the effect of EC legislation has been different in each country. The starting points may be different because the administrative structure differs from country to country, or because the existing legislation is different, or because of geographical differences.

The principal differences in administrative structure or tradition affecting the Directives under consideration are these:

- *Germany* is a federation in which responsibilities, or competences, are divided between the *Bund* and the *Laender* in different ways depending on the subject. The process of formal implementation has frequently been complicated by the fact that each *Laender* has to introduce legislation or administrative measures, and they will not all do this in the same way;
- in the *United Kingdom* nearly all the Directives under consideration

have been formally complied with by administrative circulars sent by central government to the responsible authorities. This has been possible because the existing legislation was broad enough and the Minister has the reserve power to compel the authorities to implement the measures should they fail to do so;

- in the *Netherlands* the Directives under consideration have had to be fully translated into national law, and this meticulous process has resulted in long delays in formal compliance;

- in *France* many functions are carried out at the level of the *département* by officials who are directly answerable to central government. This has simplified formal compliance, but has created difficulties with some Directives dealing with rivers which cross the boundaries of several *départements*.

The extent to which the goals of EC environmental policy have been achieved so far is a mixed story, best described as a qualified success. Some goals have been achieved while others have not, or not yet.

## GOALS ACHIEVED

Several common *standards* have been agreed in the EC. These relate to products (e.g. detergents), to emissions to water (although one country, the UK, is not in general following these standards) and water quality standards. The existing monitoring in all countries has had to be extended to ensure that these standards are achieved. Several common *procedures* have also been agreed. Thus all countries have water quality plans with pollution reduction programmes, and all are starting to introduce waste disposal plans. All waste disposal installations or sites for treating and tipping waste are being authorised.

Much greater attention has been focussed on certain dangerous substances (e.g. mercury) and on certain industries (e.g. titanium dioxide production). The setting of a standard for nitrates in drinking water has drawn attention to an increasing problem in all four countries.

## GOALS NOT ACHIEVED

It cannot be shown that the Directives have yet produced widespread improvements in water quality. The reason for this is that countries have often been slow to designate waters for the purposes of the Directives, or in some cases have not done so at all (e.g. no bathing waters in Netherlands, no waters for fish and shellfish in Germany). Indeed the most consistent failure to achieve the goals of policy has been the way in which the dates set in Directives have not been met. This can probably be explained as a failure on the part of the negotiators of Directives to realise in advance the difficulties in fulfilling the obligations that they are undertaking. These difficulties are sometimes at a national level (e.g. the introduction of new national legislation) or sometimes at the regional or local level. Some slippages in time are to be expected, and are excusable in certain circumstances, but consistent delays in implementing even the formal obligations in Directives are bound in the long run to bring the policy into disrepute.

# IMPACT ON MEMBER STATES

The greatest effect of the Directives has been to provide a stimulus to national policy making: national policies have had to be measured and justified against the yardstick of EC legislation. In the Netherlands water policy has shifted to the functional approach with standards set at national level, and with significant new legislation having to be introduced. In the United Kingdom the environmental quality objective approach for water has had to be refined. In Germany the shift to nationally set emission standards based on what is technically achievable has been encouraged by one EC Directive although in the process Germany has moved further away from the environmental quality standard approach in other EC Directives. Only in France has there been little change in water policy, although there has been a stimulus to implement the existing policy of drawing up maps of water quality objectives. In the field of waste there has been less of an impact, although the drawing up of waste and toxic waste disposal plans will be an important step when eventually it is complete.

# ADMINISTRATIVE CENTRALISATION

An unexpected effect of EC policy has been to centralise into the hands of national governments some functions that had previously been devolved to regional or local authorities. The setting of standards at EC level means that these standards are negotiated by the national government and therefore can no longer be set entirely at a local level. Furthermore the central government has to ensure that the standards are applied by local/regional authorities since it is the central government that has to report to the Commission in Brussels that the obligations in Directives has been fulfilled. This centralisation is an unintended result of EC environmental policy which can raise difficult issues particularly in a federal country like Germany.

# CONCLUSIONS

If the environmental policy of the Community is largely contained in items of Community legislation it must also include the carrying out of the obligations contained in that legislation. To understand how Community policy is implemented is therefore essential to an understanding of Community policy. Community policy is very much more than the written words in a number of texts.

This report shows that a Directive can influence national policy, national legislation, and administrative procedures as well as having practical effects. These effects vary from country to country and are not always predictable in advance. Effects are also likely to continue over time since implementation is not something that happens all at once on a date set down in a Directive. Fully to understand Community environmental policy is therefore a large task since not only does it involve an understanding of what has happened in all Member States but in each of these the effects of Directives are changing.

In a short period of years the Community has introduced a large body of Directives in the field of water and waste. Undoubtedly these have contributed to national developments and have helped to focus attention on national deficiencies. They have therefore helped to raise awareness of environmental issues. However the administrative solutions adopted for particular problems in one country are still little known in others, and despite the Directives there is still a great diversity in the different Member States. This clearly points to the need for a greater exchange of views and information between countries by the responsible authorities, at whatever level of government they may be.

# Appendix
## Chronological List of Directives

| Directive no | Reference in Official Journal | | Chapter Reference | Subject |
|---|---|---|---|---|
| 73/404 | L 347 | 17.12.73 | 3.1 | Detergents |
| 73/405 | L 347 | 17.12.73 | 3.1 | Detergents |
| 75/439 | L 194 | 25.07.75 | 5.4 | Waste oils |
| 75/440 | L 194 | 25.07.75 | 3.2 | Surface water for drinking |
| 75/442 | L 194 | 25.07.75 | 5.1 | Waste |
| 76/160 | L 31 | 05.02.76 | 3.7 | Bathing water |
| 76/403 | L 108 | 26.04.76 | 5.3 | Disposals of PCBs |
| 76/464 | L 129 | 18.05.76 | 3.8 | Dangerous substances in water |
| 78/176 | L 54 | 25.02.78 | 3.11 | Titanium dioxide |
| 78/319 | L 84 | 31.03.78 | 5.2 | Toxic waste |
| 78/659 | L 222 | 14.08.78 | 3.5 | Water standards for fresh-water fish |
| 79/869 | L 271 | 29.10.79 | 3.3 | Sampling surface water for drinking |
| 79/923 | L 281 | 10.11.79 | 3.6 | Shellfish waters |
| 80/68 | L 20 | 26.01.80 | 3.9 | Groundwater |
| 80/778 | L 229 | 30.08.80 | 3.4 | Drinking water |
| 82/176 | L 81 | 27.03.82 | 3.10 | Mercury from the chlor-alkali industry |
| 82/242 | L 109 | 22.04.82 | 3.1 | Detergents |
| 82/243 | L 109 | 22.04.82 | 3.1 | Detergents |
| 82/883 | L 378 | 31.12.82 | 3.11 | Titanium dioxide |